The Copper
Industry in the
Chilean Economy

THE WHARTON ECONOMETRIC STUDIES SERIES

Wharton Econometric Forecasting Associates and
Economics Research Unit
The University of Pennsylvania
F. Gerard Adams and Lawrence R. Klein, Coordinators

The Copper Industry in the Chilean Economy

An Econometric Analysis

Manuel Lasaga
Wharton Econometric
Forecasting Associates, Inc.

LexingtonBooks
D.C. Heath and Company
Lexington, Massachusetts
Toronto

338.2743
L33c

Library of Congress Cataloging in Publication Data

Lasaga, Manuel.
The copper industry in the Chilean economy.

Includes index.
1. Copper industry and trade—Chile—Mathematical
models. I. Title.

| HD9539.070557 | 338.2'743'0983 | 81-1439 |
| ISBN 0-669-04543-8 | | AACR2 |

Copyright © 1981 by D.C. Heath and Company

Published simultaneously in Canada

Printed in the United States of America

International Standard Book Number: 0-669-04543-8

Library of Congress Catalog Card Number: 81-1439

To my wife, Margarita, for the inspiration, motivation and dedication through all the grueling times; and to our newborn daughter, Vivian Marie, for the joy she has brought.

Contents

List of Tables

Acknowledgments

Out of the many who have contributed directly or indirectly to the fulfillment of this project, there are a few whose names deserve special mention: Professors F. Gerard Adams and Jere R. Behrman of the University of Pennsylvania, and Aldo Roldan—the first two for their guidance, criticisms, and suggestions from the beginning through the completion of my graduate-studies program in the lecture halls and elsewhere as my thesis advisors, the latter for his assistance in the specification and estimation of the econometric model (of his native country).

This project is part of a series on commodity exports and economic development that includes studies of copper and Zambia by C. Obidegwu and M. Nziramasanga, coffee and the Ivory Coast by T. Priovolos, coffee and Central America by G. Siri, coffee and Brazil by F.G. Adams and T. Priovolos, and a summary volume by F.G. Adams and J. Behrman.

I would also like to express my gratitude to the rest of the Wharton EFA-AID commodity group: D. Ford, A. Gayoso, J. Manger, C. Obidegwu, M. Nziramasanga, J. Pobukadee, T. Priovolos, G. Siri, and J. Vargas for a very fruitful association. I would also like to thank the other individuals who have contributed through their helpful suggestions during various stages of the project: W. Cline, K. Jay, F. Levy, C. Michalopoulos, T. Morrison, and L. Perez. The U.S. Agency for International Development (AID) provided financial support for this study. Work on the copper model used in this work was supported by the Rockefeller Foundation.

Finally, I would like to thank Delores White for her good work with the word-processing machine and her good attitude despite my incessant revisions.

The Copper
Industry in the
Chilean Economy

1 Introduction

Chile, copper, growth, and stabilization policies are the four protagonists of this empirical piece. Copper accounts for about two-thirds of Chilean exports and through a myriad of linkages has influenced economic growth and the formulation of stabilization policies. This theme has been prevalent in the evolution of Chilean economic thought. Among the vast literature on the Chilean economy are a number of excellent studies that have contributed significantly to the development of the methodology and empirical analysis presented in the ensuing chapters.

Chile's stake in copper is very high, and its economy has progressed despite the apparent threats of instability. The world price of copper represents the most influential exogenous variable affecting all sectors of the economy. The analysis of the interactions between the external sector through copper prices and the rest of the economy has been complicated by the dramatic political changes that occurred during the period 1960–1976. The structure of ownership of the copper industry has been transformed from that of a foreign-owned industry to one that is government-owned. Each of the mines had operated as independent foreign-owned units, but after nationalization, they were consolidated under one government agency.

The objectives of this study are to construct an econometric model of Chile that has a detailed copper sector and to evaluate the impact of copper policy variables through model simulations. Although there are numerous historical studies on this subject, there have been almost no attempts at developing the econometric analysis.[1] The advantage of the latter technique is the superiority of information revealed from model simulations. For instance, it is possible to isolate the impact of the economy from an exogenous variation in one variable, for example, copper production, which would not be feasible with partial equilibrium or data-comparisons techniques.

In specification of the model, the rigid assumptions of the Keynesian demand-determined output were modified to fit the context of a developing economy characterized by supply bottlenecks. The existence of a limited financial sector in Chile precludes the use of the standard macroeconomic theory of interest rates and the equilibrium between the goods and money markets. Thus an alternative specification was introduced that is more compatible with the structure of financial institutions in Chile. The discussion of economic theory and model specification in a developing economy such as Chile is pursued in Chapter 3.

Although there have been previous econometric models for Chile, this is a first attempt at incorporating the post-1970 period. Nevertheless, the existing models provide a vast source of ideas, some of which have been embodied in the specification of this model. Three of the more recent works are by Behrman (1977), Corbo (1974), and Lira (1974). Lira deals explicitly with the relationship between the copper and the noncopper sectors.

In *Macroeconomic Policy in a Developing Country: The Chilean Experience,* Behrman is concerned primarily with the formulation of macroeconomic policy. Through extensive analysis of model simulations, he explores the effects of alternative government fiscal, monetary, foreign-sector, income, and labor-market policies. The usefulness of these simulations is derived from the rigorous application of economic theory in the specification of a highly disaggregated model. The model is based on the Keynesian short-run macro theory with the basic distinction that output is determined from supply variables. The production sector is divided into nine subsectors that are represented by constant elasticity of substitution, or Cobb-Douglas, production functions.

The sample period for the estimation of the Behrman model is 1945–1965. The model is comprised of 172 behavioral equations and identities and 72 exogenous variables. In addition to the differences in the sample period for estimation, the Behrman model does not separate copper from any of the aggregate demand or supply variables.

Corbo's estimation of an econometric model is oriented toward a description of the dynamics of inflation in the Chilean economy. The main ingredients of the price equation are unit labor costs, price of imported raw materials, and the rate of capacity utilization. The latter variable reflects the disequilibrium in the goods market, which is influenced by disequilibrium in the money market through the consumption equation.[2] The model consists of 70 equations (30 behavioral equations and 40 identities) and 32 exogenous variables. Although the model incorporates supply constraints via the capacity-utilization variables, it is at a much more aggregate level than Behrman's model. In fact, Corbo divides the production sector into two subsectors: commodities and services. Most of the equations are estimated for the period 1960–1970.

In the specification of the model, Corbo reviews several hypotheses concerning the causes of inflation. Yet neither the theoretical analyses nor the simulations are geared specifically to the formulation of government stabilization policies. Within the structure of the model, there is no separation of copper from noncopper variables; this omits the direct link between copper and the rate of inflation.

Lira separates the copper from the noncopper sectors in a model consisting of 28 equations (11 behavioral equations and 17 identities). Aggregate output is demand-determined, and gross domestic investment is assumed exogenous. The copper sector is defined as the Gran Mineria only.[3] There are seven equations in that sector which explain the following variables: copper output, value of exports, costs of production in national and foreign currency, imports of intermediate inputs, direct taxes, and net flow of foreign exchange. The period of estimation is 1956–1968.

The work by Lira is a good attempt at examining the costs and benefits of copper in Chile within a macroeconometric model of the pre-1970 vintage. The present study extends the analysis into a more recent period and disaggregates the model into seven production sectors. The copper sector is divided into the Gran Mineria and the medium- and small-scale mines.[4] The equations for the Gran Mineria include the following variables not incorporated into the Lira model: gross investment, employment, wage determination, imports of capital, and a linkage to the other production sectors that furnish inputs to the Gran Mineria.

The material of this study is divided into five chapters. Chapter 2 summarizes the major trends in the world copper market and the concurrent Chilean policies with respect to copper and noncopper variables. This historical description deals with three distinct periods: pre-Allende, 1960–1970; Allende, 1971–1973 (nationalization); and post-Allende, 1974–1976 (military junta and economic liberalization policies). The description of the econometric model follows in chapter 3. The analysis will focus on the standard macro theory with some adaptations to the case of a developing economy characterized by supply constraints. Chapters 4, 5, and 6 deal with the model simulations. Various tests of the behavioral responses of the model are discussed in chapter 4, while the other two chapters present the copper-sector simulations that could provide the basis for the development of national copper policies. Chapter 6 deals explicitly with the issue of copper export price variability and its effect on the Chilean economy. Finally, chaper 7 extends the material in the previous chapters by presenting a feasible set of strategies for the Chilean copper industry that are compatible with the country's long-run growth and stabilization. These include a feasibile copper-development plan as well as a list of reasonable policy responses to copper market fluctuations in the short, medium, and long run.

Notes

1. The only work in this area is Lira (1974).

2. One of the explanatory variables in the consumption equation is excess supply of money.

3. The copper sector is made up of the Gran Mineria (large-scale mines), which produces 80 percent of total output, and the medium- and small-scale mines, which are heterogeneous in their technology.

4. The medium- and small-scale mines are included via a production equation, as is explained in chapter 3.

2

The World Copper Market, Chilean Copper Policies, and Economic Development, 1960-1976

This chapter presents a brief historical analysis of the Chilean economy during the period 1960-1976. Within this framework there are three elements that are essential for an understanding of the specification of the econometric model presented in the next chapter. These three aspects of Chilean economic history are the structural features of the economy, the copper industry and government policies, and the fundamental political changes that transformed the economic system. The first is essential for an understanding of the development process in Chile. In any historical analysis, it is not easy to separate the latter two issues; their importance can be observed through the interaction between politics and the copper industry.

An understanding of the linkages between the copper sector in Chile and the macroeconomic variables is imperative to the specification of an econometric model that will serve as a tool for the evaluation of alternative national copper policies. Knowledge of the historical experience provides the essential inputs for the underlying linkages in the specification of the behavioral relationships estimated in the model. The period of analysis for this specialized study encompasses seventeen years, 1960-1976; and during that time, the Chilean economy has experienced dramatic changes. In retrospect, the copper industry has contributed significantly to Chile's growth and development, while at the same time it has been subjected to fundamental structural changes through radical transformations in the country's political system. Although it had enjoyed the traditional image of a foreign-owned enclave, the industry was compelled to respond to the exigencies of a developing economy that had relied on copper for about 75 percent of its foreign-exchange earnings.

The material in this chapter is divided into two parts: a historical sketch of the principal economic features, and a brief account of the political economic events with emphasis on the copper sector. The historical sketch provides a sense of the dimensions of the Chilean economy and furnishes basic facts about the process of economic development in Chile. As an initiation to the main theme of this study, the second part lays down the groundwork for construction of the econometric model.

There are three distinct political phases in the period 1960-1976 whose differences are important to an understanding of what occurred in the economy: pre-Allende, 1960-1970; Allende, 1971-1973; and post-Allende 1974-1976. The first phase covers part of the Alessandri administration, 1959-1964, and the

5

subsequent Frei administration, 1965–1970. Since its implementation in 1955, "New Deal" legislation governed the copper sector through 1965, and in 1966, President Frei signed the Chilenization program, which sought to make the government partners in that industry. When Allende ascended to the Presidency in 1971, marking the beginning of the second phase, the country began a short-lived transition toward a centrally planned economy. One of the first, and perhaps the only, unanimously approved pieces of legislation was the nationalization of the copper industry. The political upheaval that characterized Allende's administration eventually brought about the demise of his government and his violent death. A military junta headed by General Pinochet seized control in 1973, initiating the third and current phase of development in the Chilean economy. The discussion of each phase will focus on three features: the trends in the world copper market and their implications for Chile, the Chilean economy and stabilization policies, and the principal government copper policies enacted in response to developments in the world market and to the needs of the Chilean economy.

Basic Facts about the Chilean Economy

Principal Macro Variables

As a preface to the ensuing historical discussion, the following is an account of some of the noteworthy facts on copper and Chile. In 1976, the population of Chile was 10.443 million, and the estimated GDP per capita was US$830.9.[1] Today, the structural composition of output is typical of developing economies, in that the services sector accounts for about half the total value added. However, the shares of primary and secondary sectors are less typical. In Chile, the industrial sector represents a higher share of total output and the agriculture sector represents a lower share than in many developing countries. Participation by the agricultural sector gradually diminished from 11.1 percent in 1960 to 7.6 percent in 1973. This drop reflects a significantly smaller rate of growth of real value added in this sector than in GDP. Despite the accelerated land-reform program during the Allende government, the share of agricultural value added continued to decline. From 1974 to 1976, this sector began to recuperate its share, although its share is still below the value recorded in 1960. Whether it can maintain the current upward trend will depend on the success of the military junta's economic program (see table 2–1).

The contribution of the secondary sector (which includes mining, manufacturing, construction, and public utilities) increased gradually from 37.5 percent in 1960 to 41.4 percent in 1972, which represents its highest level. As a consequence of the sharp recession in 1975, the share of the secondary sector declined to its lowest level, 36.7 percent. This deterioration in 1975 originated

Table 2-1
Chile, Composition of Real GDP by Primary, Secondary, and Tertiary Sectors,
1960-1967
(*percentage*)

Year	Primary[a]	Secondary[a]	Gran Mineria	Services[a]
1960	11.1	37.5	5.6	51.4
1961	10.8	38.4	5.8	50.8
1962	10.1	40.6	5.8	49.3
1963	10.1	40.7	5.4	49.2
1964	10.2	41.0	5.6	48.8
1965	9.5	40.7	4.7	49.8
1966	9.6	40.6	5.2	49.7
1967	10.0	40.5	5.2	49.5
1968	10.0	40.0	4.6	50.0
1969	8.8	40.9	4.7	50.3
1970	8.9	40.0	4.4	51.0
1971	8.9	41.1	3.9	50.0
1972	8.5	41.4	3.9	50.0
1973	7.6	41.4	4.5	51.5
1974	8.3	40.9	5.3	50.7
1975	9.8	36.7	4.1	53.5
1976	9.6	37.6	5.7	52.8

[a]Primary sector consists of agriculture; secondary sectory sector comprises mining, manufacturing, construction, and public utilities; and services sector includes any services such as banking, transportation, and public administration.

within the manufacturing industries, which were severely affected by the trade liberalization policies of the Pinochet administration. Manufacturing value added represents an average share of 60 and 24 percent out of the secondary sector and total value added, respectively. The period of highest growth for this subsector was between 1960 and 1970, when the development of import-substituting industries was at a peak. The average growth rates shown in table 2-2 provide some indication of which were the dynamic sectors.[2] During the pre-Allende period (1960-1970), the average growth rate for secondary sector value added was 5.1 percent, which is above the average for GDP of 4.5 percent. However, the 1975-1976 recession was responsible for the poor performance of this sector in the post-Allende period, when the average growth rate was -5.4 in contrast with the GDP rate of -2.6.

The services sector, which comprises the largest share of GDP, has exhibited dynamic growth throughout the period 1960-1976. Through the absorption of the excess supply of labor concentrated in the urban areas, it has maintained a stable share of total output of goods and services. Two key components of services are transportation and commercial activities. Because of the extensive distances between the northern and southern points of the country in contrast

Table 2-2
**Chile, Average Annual Growth Rates for Sectoral Value-Added during the
Pre-Allende, Allende, and Post-Allende Periods**
(*percentage*)

	1960–1970	*1971–1973*	*1974–1976*	*1960–1976*
Gross domestic product	4.5	–1.2	–2.6	2.7
Primary	2.3	–6.4	2.1	1.8
Secondary	5.1	–1.4	–5.4	2.7
Manufacturing	(5.4)	(–1.3)	(–8.1)	(2.1)
Gran Mineria	(1.8)	(3.4)	(–.2)	(2.7)
Services	4.4	–.2	–1.3	2.8

with its narrow width, transportation services constitute a crucial sector, and the annual average growth rate of this sector of 5 percent seems to have been sufficient to sustain the needs of a growing industrial sector. The largest share of services is provided by commercial activities, which range from the big import-export firms and commercial establishments to the street vendors, and this has been one of the less dynamic sectors of the economy.

Total output of the economy, that is, the total value added for the primary, secondary, and tertiary sectors, has shown a cyclical behavior that can be identified with the different political regimes. The annual percentage change in GDP, as well as the annual percentage change in the GDP price deflator, gross domestic investment, and the London Metals Exchange (LME) price of copper, is presented in table 2-3. This depicts the cyclical path of domestic activity, the period averages of which were presented in table 2-2. The changes characterizing the three periods are evinced by the fluctuations in GDP. The highest growth year is 1971, with a 7.7 percent increase. At the end of 1970, the economy had sufficient excess capacity to enable the Allende government to induce an expansion in domestic production (as occurred in 1971) to meet the swelling demand aroused by the income-redistribution and price-controls programs. The maximum percentage variation in GDP for the whole period is a drop of 11.3 percent as a result of the 1975 recession during the Pinochet administration.

Gross domestic investment, as seen in table 2-3, has not been a particularly important element in inducing more dynamism into the economy. Because of the small changes in net investment, the quality of the overall stock of capital has not improved significantly. In 4 out of the last 5 years, gross domestic investment has declined.

One of the central themes of Chilean economic history has been the persistent and erratic inflation and the sometimes futile attempts by the government to stabilize the price level. The minimum rate of inflation was recorded in 1971 (7.4 percent), and the maximum, in 1974 (654 percent). Each of the three regimes exploits the policy instruments with substantially different results.

Table 2-3

Chile, Annual Percent Change of Selected Economic Indicators, 1960–1967
(*percentage*)

Year	GDP (1965 pesos)	GDP Deflator (1965 = 100)	Gross Investment (1965 pesos)	LME Copper Price (U.S. ¢/lb)
1961	6.1	7.4	18.9	–6.9
1962	4.6	15.7	3.1	2.1
1963	5.1	43.5	7.8	–0.1
1964	4.3	46.5	–3.4	49.6
1965	5.0	34.9	4.5	33.1
1966	7.0	30.3	1.4	18.8
1967	2.4	28.3	1.4	–26.3
1968	3.0	30.9	7.6	10.0
1969	3.5	40.6	4.6	17.9
1970	3.6	38.8	2.8	–3.2
1971	7.7	23.6	1.1	–23.2
1972	–0.1	85.4	–15.3	–1.4
1973	–3.6	426.7	–2.8	66.3
1974	5.6	653.7	13.9	15.6
1975	–11.3	391.1	–27.1	–40.0
1976	4.1	234.7	–4.6	14.1

Whereas the rate of inflation, as measured by the percentage change in the GDP price deflator, has fluctuated with rates known only to few countries, its variability is overshadowed by that of the LME price of copper. The variability index of copper price is 3.4, while for the inflation rate and GDP growth it is 1.4 and 1.6, respectively.[3] The marked contrast between the variability of copper price and the domestic activity variables is indicative of the ability to stabilize the domestic economy from fluctuations in the external sector. A comparison of the percentage changes in copper price with those of the GDP price deflator discloses a very small correlation between them. For certain other countries, however, especially basic agricultural commodity exporters, the correlation comes out as statistically significant.

The Copper Industry

The development of the Chilean economy between 1960 and 1976 has been affected by the evolution of the copper industry. The purpose of this study is to identify those linkages and to quantify the extent of their impact in order to suggest some policy alternatives. In view of this objective, some of the important economic and institutional characteristics of the copper sector are also summarized in this section. Within the Chilean copper industry there is a dichotomy between what is called the Gran Mineria and the medium- and small-scale mines

(Pequeña y Mediana Mineria). The former has contributed, on average, 83 percent of annual production of copper, and because of the availability of data as well as the relatively homogeneous structure of production units, most of the econometric analysis in this study will deal with the role of the Gran Mineria in the development of the Chilean economy. The medium- and small-scale mines comprise a heterogeneous group of a few capital-intensive mines and numerous small ones that in some cases are exploited by a few workers using the most rudimentary technology: pick and shovel.

Before the implementation of the nationalization policies, the Gran Mineria was made up of five mines: El Teniente (owned by Kennecott), Chuquicamata, Potrerillos, El Salvador, and Exotica (all owned by Anaconda). The El Teniente and El Salvador mines are underground. As part of the Chilenization program during the Frei administration, another mine, the Andina, was developed in 1966. During the post-Allende period, the average share of total output contributed by each of the mines was as follows: Chuquicamata, 49.5 percent; El Teniente, 31.7 percent; El Salvador, 10.6 percent; and Andina, 8.2 percent. Both Exotica and Potrerillos mines had been closed. Total copper production by the Gran Mineria and the medium- and small-scale mines and Chile's share out of total world production between 1961 and 1976 are presented in table 2-4. With

Table 2-4
Copper Production: Gran Mineria, Medium- and Small-Scale Mines, and World Total, 1961-1976
(000 M.T.)

Year	Gran Mineria	Medium- and Small-Scale Mines	World	Chile's Share[a]
1961	481.1	66.7	4335.5	12.6
1962	510.3	75.8	4491.2	13.0
1963	570.4	93.7	4559.0	13.2
1964	527.8	93.9	4764.7	13.0
1965	479.2	106.1	4968.6	11.8
1966	536.8	99.9	5205.7	12.2
1967	536.4	123.8	5059.0	13.0
1968	519.3	137.7	5457.8	12.0
1969	540.2	147.9	5953.4	11.6
1970	540.7	150.9	6382.3	10.8
1971	571.3	137.0	6449.7	11.0
1972	592.3	124.2	7050.6	10.2
1973	615.3	120.1	7510.1	9.8
1974	762.3	139.2	7659.3	11.8
1975	682.3	146.0	7338.1	11.3
1976	854.1	151.1	7871.0	12.8

[a]Percentage share of total Chilean copper production out of total world production.

the exception of 1973, Chile's share of world production has stayed around 12 percent.

Chile's copper reserves are estimated at 150 million metric tons of metallic copper contained in some 23 billion metric tons of ore averaging 0.66 percent Cu.[4] Proven reserves are 49.9 million metric tons of metallic copper contained in some 4.1 billion metric tons of ore averaging 1.23 percent Cu. The high grade of proven reserves explains why Chilean mines operate under one of the lowest operating costs in the world. At the current rate of production of 1 million metric tons per year, proven reserves will last about 50 years. Total smelting capacity in 1973 was estimated at 805,700 metric tons, so that the rate of capacity utilization would have been 91.3 percent.

The principal direct linkages between the Gran Mineria and the rest of the economy are through the balance of payments and the public-sector budget. Tables 2-5 and 2-6 show the size of the Gran Mineria relative to four domestic-activity variables: GDP, exports, employment, and central-government revenues.[5] The average share of the Gran Mineria with respect to GDP is 5.0 percent, which represents a very modest share of total domestic output. However, the share of total exports is much higher (see column 2 of table 2-5). Chile has been able to diversify its production activities, especially the industrial sector, avoiding the dual economy structure characteristic of other primary commodity-producing countries. In addition, the Gran Mineria accounts for barely 1 percent of total employment. The very limited employment opportunities are a consequence of the industry's small size relative to GDP and to the use of capital-intensive technology. From Table 2-6 it is seen that, on average, copper taxes have represented about 83 percent of the central government's current U.S. dollar revenues and about 11 percent of total revenues. The most significant change was recorded during the Allende period (1971-1973). The sudden drop in the share of copper taxes out of total revenues was brought about by the nationalization: first through a change in definitions, so that some copper revenues would be counted as income from public firms, not as tax revenues, and second through the marked decline in before-tax earnings that resulted from the difficult transition period.

The most pronounced direct impact on the Chilean economy from the copper sector is through the balance of payments. Columns 2 and 3 in Table 2-5 show the share of Gran Mineria and total copper exports out of total exports of goods in U.S. dollars. The average shares for the Gran Mineria and total copper exports between 1960 and 1976 were 57 and 70 percent, respectively. As the subsequent econometric analysis will demonstrate, export earnings are a crucial link with the macro model. During the pre-Allende period, a part of those exports earnings were returned to the Chilean economy and the remaining portion was remitted to the foreign owners. Since nationalization, most of these earnings have been retained.

Table 2-5

Chile, Selected Indicators for Gran Mineria Value Added, Exports, and
Employment, 1960–1967
(*percentage*)

Year	GM/GDP[a]	Gran Mineria: Total Exports[b]	Copper: Total Exports[c]	Gran Mineria: Total Employment[d]
1960	5.6	61.7	68.4	.8
1961	5.8	57.5	65.7	.8
1962	5.8	56.6	66.1	.8
1963	5.4	56.5	67.4	.8
1964	5.6	48.1	61.4	.8
1965	4.7	44.9	62.7	.8
1966	5.2	55.2	69.1	.8
1967	5.2	59.7	74.6	.8
1968	4.6	58.9	75.1	.8
1969	4.7	61.4	78.9	.8
1970	4.4	59.5	75.5	.8
1971	3.9	57.8	70.3	.9
1972	3.9	58.5	72.9	.9
1973	4.5	66.7	80.6	.9
1974	5.3	65.8	76.6	1.0
1975	4.0	44.9	55.3	1.1
1976	5.7	50.5	59.8	1.1

[a]Share of Gran Mineria out of total value added.

[b]Share of Gran Mineria out of total exports (millions of U.S. dollars).

[c]Share of copper out of total exports (millions of U.S. dollars).

[d]Share of Gran Mineria out of total employment.

Government Policies and the Copper Sector
during the Three Political Periods

One of the outstanding features of the Chilean economy between 1960 and
1976 was the transformation of the political system in 1971 and in 1973. Before
1971, one set of rules were followed that had been implanted since the end of
World War II; between 1971 and 1973, these rules were drastically altered by
Marxist ideas; and since the end of 1973, some of Adam Smith's doctrines have
become the focal point of the administration's economic policies. The three
phases are pre-Allende, 1960–1970; Allende, 1971–1973, and post-Allende,
1974–1976.

In Chile, the pre-Allende regime was characterized by the traditional
protection-oriented economy. Its origins can be traced to the period following
the Great Depression in the United States, when copper became an important
source of foreign exchange to the economy. Import-substitution policies created
high rates of protection and excess capacity. Throughout the period, economic
policy was primarily oriented toward eliminating the persistent inflation. During

Table 2-6
Chile, Share of Copper Taxes out of Total Current Revenues of the
Central Government, 1960-1976
(*percentage*)

Year	Copper Taxes per Total US$ Revenues[a]	Copper Taxes per Total Revenues[b]
1960	79.6	13.0
1961	75.3	9.4
1962	89.4	11.1
1963	86.8	11.9
1964	85.5	12.2
1965	85.4	11.4
1966	90.9	16.0
1967	87.8	14.3
1968	88.4	13.3
1969	90.2	15.2
1970	91.2	16.1
1971	64.1	2.0
1972	73.7	1.3
1973	67.2	1.0
1974	88.1	8.1
1975	80.7	9.7
1976	91.9	15.2

[a]Copper taxes in millions of U.S. dollars as a share of total current U.S. dollar revenues of the central government.
[b]Copper taxes in millions of pesos as a share of total current revenues of the central government (foreign and domestic currency). All U.S. dollar values were converted to pesos using the official commercial banks' exchange rate.

the Frei administration (1965-1970), emphasis was placed on greater participation of the government in the economy. This is brought out in the Chilenization of the copper sector and the agrarian reforms program, which would have gradually given the government control over operation of the former and distribution of the ownership of the latter.

In 1971, President Allende sought to establish a democratic-socialistic government. The economic implications of these changes were state ownership of the productive sectors and a sizable redistribution of income. This period was short-lived, since Allende was overthrown in 1973 and a military government was established. In spite of the short span of time, the Chilean economy experienced abrupt changes accompanied by its worst inflation in many years. The outcome of this unsuccessful experiment with socialism was a stiff reaction by the new administration, whose first objective was eliminating the excessive inflation at whatever the cost.

After the military coup in September of 1973, the government of General Pinochet adopted free-market policies and rigid monetary restraints. One of the primary objectives was the elimination of the complicated system of controls that had evolved during the previous three decades. These included the simplification of the tariff system and elimination of controls over interest rates. These measures differ from the previous two regimes in their rejection of socialistic economics and the traditional system of overprotection with excessive use of controls. Liberalization schemes for trade, prices, and interest rates were part of a new policy package to stabilize the price level. Other measures adopted by the military government were the return of nationalized companies to the private sector and the passage of legislation that would favor the foreign investor and dispel fears of expropriation.

The Pre-Allende Period: 1960–1970

The World Copper Market. In the period 1960–1970, the world price of copper, as represented by the LME price, was almost constant during the first 4 years, it increased to record levels during the following 3 years, and the pnenomenon was repeated during the last 3 years of the decade. Thus 1964–1966 and 1968–1970 were exceptionally good years in the world copper market. U.S. military involvement in Vietnam was one of the causes of the upswings in the price of copper.[6] Two important features of the world market during this period were the existence of a two-tiered price system and the formation of CIPEC (Intergovernmental Council of Copper Export Countries).

The so-called two-price system is derived from the situation prevalent in the 1954–1956 and 1964–1970 periods, during which the U.S. producers set their price below the market price.[7] The LME price is considered the market price because of the large number of contracts that specify an LME price on delivery, even though the actual physical volume transacted on the floor of the exchange is very small compared with the world volume. The second price is the major primary producers' price, which reflect trends in production costs with little response to short-run demand fluctuations. Whenever the two-price system was operating, the major producers would resort to rationing of the available supplies of copper. By introducing this aberration in the world market, producers argued that they could reduce the possibility of long-run substitution of aluminum for copper. McNicol (1975) adds another consideration, which is the profitability to partially integrated producers from quantity discrimination. Although he finds evidence supporting these contentions, he questions the general validity of his conclusions. Since the Gran Mineria was owned by major producers in the United States, its exports of copper were tied to the U.S. producers' price until the initiation of the Chilenization program toward the end of 1966.

The formation of CIPEC in 1967 marked the creation of another potentially disruptive force in the world market. CIPEC membership consisted of Chile, Zambia, Peru, and Zaire, and they accounted for about 40 percent of the non-communist world's copper production. Their objectives were to follow as much as possible the market strategies pursued very successfully by their counterparts in the Organization of Petroleum Export Countries (OPEC). The first concerted effort by CIPEC to affect the world copper price was in 1974, when the organization called on its members to reduce their exports by 15 percent.[8] However, the announced policy had little impact on the market because of the reluctance of its members to curtail their exports in view of the risks involved in expecting other members to cooperate and the threat of long-run demand substitution for copper.

The Alessandri and Frei Administrations. Both the Alessandri and the Frei administrations attempted to stabilize domestic inflation through austerity measures, but their successes were short-lived.[9] After the first year of anti-inflationary policies pursued by the Alessandri government, the inflation rate sank to an extraordinary 7.4 percent in 1961, followed by a moderate rise to 15.7 percent in 1962. The policy package included a liberalization of import tariffs, reduction in the growth of the money supply, and elimination of wage and price controls. In order to increase public-sector investment without threatening the rate of inflation, extensive use was made of foreign loans, which would also finance the increase in imports. Accompanying the trade-liberalization program was a fixed exchange-rate policy of 0.00105 pesos per U.S. dollar. The sudden success of these policies was followed by an unexpected reversal in 1963 to the previous unstable price situation. The inability of the Alessandri government to hold down the public-sector deficit coupled with too many imports, which were rapidly drawing down available reserves, led to a devaluation of the peso at the end of 1962 by more than 55 percent. The recurrent instability spread instantly, and the economy began to register rates of 43.5 and 46.5 percent inflation in 1963 and 1964.[10]

The stabilization program of the Frei administration was somewhat successful in reducing the rate of inflation during 1966–1968, but not by as much as the previous administration. In his description of the administration's plans, Ramos (1977) points to an emphasis on eliminating inflation through a reduction in costs and in inflationary expectations. These measures were implemented through the wage and price adjustments mechanism. However, the failure by the public sector to control accelerating deficits resulted in a renewed inflationary spiral.

Chilean Copper Policies, 1960–1970. Since the end of World War II, the goal of government policies had been to gradually increase Chile's participation in the earnings from the Gran Mineria.[11] This was accomplished through greater taxes and an expansion in the ownership of the mines. The implementation of these policies in the pre-Allende period was made possible with the passage of two major pieces of legislation: the New Deal (Nuevo Trato) laws passed in 1955, and the Chilenization program in 1966. The New Deal laws were aimed at revising the tax formula, establishing rules on the return of foreign exchange, and expanding the participation of domestic industries in the production activities of the Gran Mineria. The organizational structure was established through the creation of the Copper Department (Dep. tamento del Cobre). This agency was to be responsible for the collection of tax revenues, the supervision of purchases by the Gran Mineria of domestic and foreign inputs, and the formulation of marketing strategies. The tax laws eliminated the surcharge on the realized copper price in favor of a flat 50 percent income tax and a surcharge whose rate would vary inversely with production. The New Deal regulations were aimed at intensifying production, since it was believed that previous tax laws had discriminated against copper to the extent of creating stagnation in that sector.

The tax and supervisory regulations entailed within the New Deal legislation represented the principal prenationalization attempt to increase Chile's share of the foreign-exchange earnings. The two sources of foreign exchange were the income taxes paid by the foreign-owned companies in U.S. dollars and the purchase of pesos from the Central Bank for payment to suppliers of domestic inputs (materials, capital, and labor). Thus, by exercising some control over the amount of domestic inputs purchased by the Gran Mineria, the Central Bank would obtain added foreign exchange, and domestic producers would reap more business. An alternative to the supervisory mechanism was to apply a special exchange rate for purchases of pesos by the Gran Mineria, which, if kept sufficiently below the official exchange rate, would enhance the Central Bank's holdings of foreign exchange. The drawback to this approach was that higher domestic costs would bring about substitution by foreign inputs and adversely affect production.[12] This is one of the policy issues to be analyzed in chapter 5 on copper-sector simulations. The success of the domestic-inputs policy is evident from the evolution of production costs for the Gran Mineria presented in table 2-7. In 1960, the shares for domestic and foreign inputs were 27.5 and 23.6 percent, respectively, while in 1976, these were 54.4 and 14.2 percent. The Copper Department, which was responsible for this program, had issued two rules of operation: one was to grant permits for importing those inputs which could not be found in the local market at a comparable price and quality, and the other was the making of advance notices by the copper companies of future input requirements so that domestic firms would have the time to consider the feasibility of producing those inputs. These regulations were applicable for material inputs only; after 1965, some purchases of capital equipment were subjected to control. Yet, as shown in table 2-8, the large capital requirements of the Gran Mineria operations would have offered much potential to the domestic industry.

The Chilenization program went one step further than previous schemes by making the Chilean government owners of 51 percent interest in the Gran

Table 2-7
Chile, Structure of the Costs of Production for the Gran Mineria, 1960-1976
(*percentage*)

Year	Wage Bill	Domestic Inputs	Foreign Inputs	Total
1960	48.9	27.5	23.6	100.0
1961	64.3	17.9	17.8	100.0
1962	60.1	23.3	16.6	100.0
1963	55.0	24.8	20.2	100.0
1964	58.7	21.0	20.3	100.0
1965	57.6	22.2	20.2	100.0
1966	53.6	30.6	15.8	100.0
1967	67.5	17.2	15.3	100.0
1968	61.3	23.2	15.5	100.0
1969	57.0	29.2	13.8	100.0
1970	56.7	30.8	12.5	100.0
1971	66.4	23.6	10.0	100.0
1972	61.0	36.3	2.7	100.0
1973	41.6	49.6	8.8	100.0
1974	25.4	56.6	18.0	100.0
1975	27.5	42.8	29.7	100.0
1976	31.4	54.4	14.2	100.0

Table 2-8
Chile, Capital-Labor Ratios for the Gran Mineria and for the Economy
(Excluding Gran Mineria), 1960-1976
(*thousands of 1965 pesos per person*)

Year	Non-Gran Mineria	Gran Mineria
1960	21.3	155.7
1961	21.5	140.4
1962	21.4	142.2
1963	21.4	133.6
1964	21.3	124.0
1965	21.1	116.7
1966	21.1	114.7
1967	20.7	109.1
1968	20.7	115.8
1969	20.9	125.8
1970	20.9	126.3
1971	20.4	114.6
1972	20.0	97.7
1973	20.2	86.6
1974	20.9	93.1
1975	21.6	102.6
1976	20.9	105.7

Mineria. These arrangements created a new source of income via dividends and assigned to the government the principal voice in the operating decisions of the mines. As of December 1970, the government had acquired the following shares in each of the mines: 49 percent in Chuquicamata, 49 percent in El Salvador, 40 percent in El Terriente, 75 percent in the Exotica, and 70 percent in the Andina. Both the Andina and the Exotica mines had begun operations in 1970. The government also approved an investment program that would significantly increase production beginning at the end of 1970.

The Allende Period: 1971-1973

The World Copper Market. In the first year of the Allende administration, the LME price of copper plunged 23 percent; it then remained steady for 2 years; and in 1973, it climbed to 66 percent. Since Chilean exports had been priced on the LME, these fluctuations were immediately reflected in export earnings. The variations in the world price for copper were a response to a leveling-off in world demand between 1969 and 1971 and the boom in 1972 and 1973 that originated in the United States, Japan, and West Germany. The major consideration of CIPEC during this period was the consolidation of its membership and the support of the Chilean government in its struggle against the U.S. companies' opposition to the terms of the nationalization policies.

The Allende Administration. While the inflation rate was moderately low during the first year (23.6 percent), the instability of the Allende government and the incompatibility of economic policies resulted in hyperinflation in 1973, with an increase of 427 percent in the GDP price deflator. The precipitous rise in the price level is explained by Ramos (1977) as repressed inflation. At first the administration embarked on an ambitious income-redistribution program that would have been achieved, in part, through the nationalization of large private companies. Fortunately, the build up of excess productive capacity prior to 1971 and the existence of ample supplies of foreign reserves enabled the economy to meet the strong rise in demand. The year of highest growth in GDP between 1960 and 1976 was 1971, with 7.7 percent, and the inflation rate that year was the lowest recorded in the previous 10 years. The impressive success registered during the first year appears to be somewhat fictitious, since it was based on the official price indices. The Allende government had imposed price controls on many of the items in the consumer market. The underestimation of actual inflation as derived from the use of the official price indices is evident from the widespread use of black markets in which prices were far above the official price.[13] Ramos (1977) adds another explanation for the repressed inflation phenomenon: the potential inflationary impacts of the large government deficits in 1971-1972 were prevented in view of a significant increase in demand for money induced by fears of expropriation. Thus the existence of excess capacity, availability of reserves, price controls, and the increased demand for money proved to be a temporary relief from the inherent disequilibrium that eventually brought down the government and introduced into the economy the onerous experience of hyperinflation.

One of the principal objectives of the Allende plan was a significant redistribution of income. The chief instruments for the redistribution of the national wealth were (1) much higher salaries, (2) more employment by the public sector, (3) a breakup of monopolies through nationalization, (4) price controls on basic items in the consumer markets, and (5) greater transfer payments to individuals financed by earnings from the nationalized industries. The immediate success of the Allende administration in improving the distribution of income is evident from the period averages for the share of wage out of total income, which between 1960 and 1970 was 48.7 percent. In 1971 and 1972, the share of wage income was 61.7 and 62.8 percent, respectively, in contrast with the pre-Allende period, during which the same share did not go much beyond 50 percent.

Chilean Copper Policies, 1971-1973. In July of 1971 the Congress approved the constitutional amendment nationalizing the copper industry. It was to be the only instance in which the Congress and the President would arrive at a concensus on a legislative issue. The nationalization law transferred ownership of the mines and the refineries to the government, which was then responsible for the determination of adequate compensation to the previous owners. As the spokesman for the government, President Allende was to determine the book value of the companies and apply a deduction for any "excess profits" earned since 1955.[14] Although the government had ownership of about 50 percent of the mines before the nationalization, these shares had been acquired through longterm credit agreements with the companies and were also subject to the terms of the compensation formula. After the deduction for excess profits, the government announced that no compensation was due the companies.

The immediate reaction by Kennecott was to seek an embargo on Chilean exports of copper through the international courts. Initially it succeeded in preventing sales to some European countries; however, these did not include the largest consumers: Germany, Britain, and Italy. Afterwards a decision by the German courts ruled in favor of the Chilean government's right to continue their sales. Kennecott resorted to other tactics, such as pursuing a U.S. court injunction to prevent the sales of materials and equipment to Chile. The anti-Allende campaign indicated by Kennecott and supported by the U.S. government may not have persisted all the way to the downfall of the government had Allende agreed on some minimum compensation, that is, nationalization with compensation. Even a partial compensation for the mines could have allayed the powerful opposition to the Allende government by the United States and other industrial nations.

The administration of the mines was assigned to CODELCO (Corporation del Cobre), which had replaced the Copper Department in 1966 as part of the Chilenization program. The difficulties encountered by CODELCO were mostly of an organizational nature derived from its inexperience with the daily operations of a complex industry. The transition to a nationalized copper industry also was made difficult by political circumstances. The apprehension of political instability accompanied by generous offers from the U.S. copper companies to transfer abroad led to the flight of high-level technicians. Because of the actions taken by Kennecott to stop shipments of materials and spare parts, the copper mines were unable to maintain previous high rates of production.

The Allende administration's income and production tax policies were basically unchanged from the past. However, in 1972, the government announced a special exchange rate for the Gran Mineria significantly below the official pesos/dollar rate. This exchange rate applied to any purchases of pesos by the copper companies for payments to domestic inputs. The overvalued peso rate was maintained through 1973, and in the last quarter of 1974, the Pinochet government decreed a unified exchange rate for all foreign-exchange transactions: the official commercial bank rate.[15] Between 1972 and 1974, the foreign-exchange-rate policy increased the availability of foreign exchange to the central bank while penalizing the copper industry through higher costs. The economic benefits of this policy are not apparent, since the government, as owner of the copper industry, was actually penalizing its own income-generating activities by imposing an overvalued rate for all domestic transactions.

The nationalization of the copper industry during the Allende administration introduced two additional policy considerations. One was the establishment of a new source of income. The remittances of net profit, either in the form of dividends or retained earnings, that were being channeled to the foreign owners now would flow into the coffers of the state. The other was the realization by the government of its responsibility as the owner of the copper industry to define a copper policy within the context of the long-run development objectives of the Chilean economy.

The Post-Allende Period: 1974–1976

The World Copper Market. The world copper market during this period was severely affected by the world recession. In 1974, the LME price of copper increased by 16 percent, reaching the highest level recorded for that market at 93.4 US cents/lb. However, the world recession of 1974–1975 was quite severe, and this was reflected in a drop of 40 percent in the LME price. The recuperation of the world economies that began in 1976 strengthened demand for copper and lifted the LME price by about 14 percent. Thus the sharp decline and swift recovery of the LME price coincided with the strong recession and rapid recovery of the Chilean economy from 1974 to 1976.

World copper production was stagnant between 1973 and 1976. The unsuccessful attempt in 1974 by CIPEC to curtail production does not explain the feeble performance by world producers. The latter was more likely a response to the recession in the copper-consuming countries. Total output increased by 3 percent during that period in contrast with an increase in Chilean production of 37 percent. Chile's share of the world market rose from 9.8 percent in 1973 to 12.8 percent in 1976.

The Pinochet Administration. The economic philosophy of the Pinochet government was to promote a free-market system by eliminating unnecessary controls and reducing state ownership of the means of production. The chronology of events from 1974 through 1976 depicts the step-by-step deregulation of some of the more important price-determining mechanisms. Arbitrary rules were to be replaced by the interaction of supply and demand in the marketplace. Each of these measures was as dramatic or even more so than the policies adopted by Allende.

One of the first areas affected by the new policies was the foreign-trade sector. The development of the import-substituting industries in the previous 25 years had brought about the establishment of a complicated import tariff code. There was an almost interminable list of items with their corresponding rates. Different exchange rates also were used to discriminate against certain luxury goods or favor essential material imports. The costs to the government of administering and enforcing these codes were formidable. In reaction to this state of affairs, the Pinochet administration announced that a uniform tariff rate would be applied to imports after a gradual dismantling of the then current system.[16] The timetable for this program projected that the average tariff rate on all imports would be reduced from 94 percent in the second semester of 1973 to 20 percent in the second semester of 1977.[17]

The top priority of the military government was the elimination of hyperinflation. In 1974, Chile had experienced the highest rate of inflation ever, with an increase in the GDP deflator of 654 percent.[18] The biggest challenge facing the government was what Ramos (1977) describes as the repressed inflation and the control of price expectations. From the end of 1973 through 1975, the peso was devalued considerably in order to reestablish a stable purchasing-power parity rate. The average yearly rates of devaluation were 469, 651, and 490 percent in 1973, 1974, and 1975, respectively. Along with the strong devaluations, the junta adopted a rigid wage policy. The wage adjustments that were due in October of 1973 were postponed until March of 1974. In order to prevent the workers from expressing their discontent, the government banned the right to strike.

An innovative feature of this administration's price-stabilization program was the elimination of almost all existing price controls. Its purpose was to prevent supply bottlenecks and the spread of black markets. If the government was successful in bringing down expectations, then the market price would be able to adjust less violently to any short-run disequilibrium. At first, prices increased much more than expected, so demand fell abruptly, precipitating the recession in 1975. However, the success in bringing down inflation has been impressive. The price liberalizations also applied to interest rates, so for the first time in decades firms faced real rates of interests on their loans. This extraordinary increase in the cost of borrowing affected the financial stability of many firms, and during the recession, many had to declare bankruptcy.

The year in which the LME price of copper dropped by 40 percent, the Chilean economy experienced a severe recession. Even though the two events coincided, the drop of 11 percent in the GDP during 1975 was a consequence of the rigid economic policies of the Pinochet administration. The breakdown of trade barriers had exposed many Chilean companies to sharp competition from imports. The very high interest rates meant sizable increases in operating costs to most producers. Finally, firms had set prices too high in anticipation of greater inflation; thus when real incomes declined as a result of tight wage policies, demand fell far short of supply.

The undoing of Allende's legacy was one of the priorities of the military government. The unsuccessful experiment with socialism left the public sector with the ownership of many firms distributed throughout all sectors of the economy. With the exception of copper (CODELCO is the largest company in Chile) and some of the strategic industries (electricity, telephones, and so forth), the government opened bidding on a large number of companies or made special arrangements with the previous owners.[19] In order to attract foreign capital in joint partnerships, the foreign-investment code was revised. Because of its leniency toward foreign investors, the code was unacceptable to the Andean Pact and Chile withdrew from that organization.

Chilean Copper Policies, 1974-1976. The success of the government in stabilizing domestic prices was matched by the success of the copper industry in reducing production costs. The combination of exchange-rate policy and CODELCO's management problems had increased substantially the U.S. dollar cost per pound of production during the Allende period. In 1973, the average cost of production in the Gran Mineria increased by 43 percent to the highest level ever recorded. A massive cost-control program was able to reduce that cost by 51 percent between 1973 and 1976, According to Copper Studies, Inc. (1978), the four reasons for the reduction in costs were (1) a significant increase in production, (2) the closure of the Exotica mine in 1975, (3) the wage and labor relations policy of the military government, and (4) the exchange-rate policy.[20]

Because of the abundance of Chile's copper reserves, the Pinochet government sought to stimulate production in that sector. Between 1974 and 1978, 90 percent of direct foreign investment approved by the government was in the mining sector.[21] The Gran Mineria accounted for a very large share of that total.

The high grade of ore in the proven reserves implies that Chile can maintain its very low operating costs. Because of its comparative advantage, copper will thus remain as the most important source of foreign-exchange earnings. If the current levels of planned investment are realized, the expansion in copper production could significantly increase Chile's share of the world market. Since the Gran Mineria produces 80 percent of the industry's output, the larger share of the world's market would enhance the governments monopoly power in the sale of copper.

Notes

1. This calculation is based on DeCastro (1978).

2. A reasonable definition of a dynamic sector is one whose average growth rate exceeds that of the gross domestic product (GDP).

3. The variability measure is based on the percentage changes for the three variables shown in table 2-3; it is defined as the sample standard deviation divided by the corresponding mean.

4. Most of the data on Chile's endowments and smelting capacity are derived from *Engineering and Mining Journal* (1973,1977).

5. Data on copper taxes were available only in the current account of the central government. The latter does not include public firms, decentralized agencies, and municipalities, so any conclusions based on central government data must acknowledge these omissions.

6. For more discussion of the world copper market and its evolution, see Banks (1974).

7. For a good analysis of the economic rationale for the use of the two-price system between 1949 and 1966, see McNicol (1975).

8. See Zorn (1978).

9. The analysis of stabilization schemes in Chile has been a prevalent topic in the economic literature of that country. It is not an economic problem that has characterized one particular period; it has plagued the development of Chile since the last century. This chapter attempts to summarize very briefly the policies that were enacted and their general effects. An excellent literature exists on this subject matter; some of the more notable ones are Ffrench-Davis (1973), Arbildua and Luders (1968), Ramos (1977), and Luders (1970).

10. See table 2-3.

11. There are numerous sources dealing with copper and the Chilean economy. These are mostly historical analyses of government policies: Ffrench-Davis (1974*a*), Tironi (1974), Barria (1974), Ffrench-Davis (1974*b*), Tironi (1977). The only attempt at applying an econometric model to analyze policy issues is Lira (1974) for the period 1956-1968.

12. Ffrench-Davis (1975), in an empirical evaluation of the import-substitution issue, estimates the price elasticity of demand for domestic relative to foreign inputs as -0.6 based on data for 1955-1970.

13. This is cited by various authors including Ramos (1977) and Mamalakis (1976).

14. See Faundez (1978) and Fortin (1978) for a description of the legal process involved in the nationalization of the Gran Mineria.

15. The Gran Mineria exchange rate was lower than the official commercial bank rate by 10, 52, and 20 percent in 1972, 1973, and 1974, respectively. This information was obtained from COLDELCO (1974).

16. There would still be some exceptions, such as automobiles, but these would be kept to a minimum.

17. According to the Central Bank monthly bulletin, these goals have been achieved.

18. This would be the official rate; however, one study by the Universidad Catolica has revealed a 30 percent underestimation by the official price indices.

19. Despite the efforts to desocialize the economy, in 1977, of the top 100 companies (ranked by sales), 25 were owned or controlled by the state. These companies represented 77 percent of total net worth. Other interesting facts concerning the top 100 are found in the December 1978 issue of *Chile Economic News* published by CORFO.

20. As was mentioned earlier in this chapter, the Gran Mineria exchange rate was the same as the official rate after October 1974. The Exotica mine had been a high-cost operation, so its closure reduced the average production cost of the mines.

21. See *Chile Economic News,* January 1979.

3

Description of the Econometric Model

The importance of copper to the Chilean economy has been amply documented in terms of a historical analysis of the more relevant structural variables. However, these studies have concentrated on the partial one-way linkages between copper and the rest of the economy. The objective of this work is to test some of the hypotheses concerning the interactions between copper and the other sectors through the estimation of an econometric model. This approach reveals many of the dynamic two-way linkages in great detail and makes possible the evaluation of the impacts of alternative copper policies on production, income, and distribution of copper, as well as on noncopper variables, within a single period and in an intertemporal context. The identification of the structure of the model and the discussion of the principal behavioral equations will be dealt with in this chapter.

One of the main objectives of the Chilean model is to trace all possible linkages between the copper sector and the rest of the economy. Since production of copper by the Gran Mineria accounts for about 80 percent of the total, most of the equations relating the copper sector to the domestic economy refer to the Gran Mineria. The direct contribution of copper to the gross domestic product (GDP) is not very large. However, the indirect effects are very important, and the ability to capture these links has been a first priority in the specification of the model.

The direct linkages of copper to the rest of the economy are through its value-added components: employment, wage bill, return to other factors, and so forth. The indirect linkages operate through the balance-of-payments variables (exports, ability to import, and exchange rate), government-sector variables (current revenues, expenditures, and savings) demand variables (consumption and investment), and the determination of value added for those sectors providing inputs into copper production.

The role of copper in the development of the Chilean economy during the 1960–1976 period has been elaborated from a historical perspective in the previous chapter. Yet few tenable conclusions about the impact of national copper policies enacted during that period can be derived from a causal analysis of those figures without first isolating the concomitant effects of other variables. The use of a disaggregated econometric model that incorporates a detailed copper micro sector with ample feedbacks to the other sectors is an attempt to grapple more realistically with these issues. The material presented in this chapter represents the culmination of a search process whereby for each equation numerous hypotheses were tested before arriving at that relationship which best described the Chilean economy during that period. How realistic the assumptions and behavioral properties of the model are will be dealt with in the following chapter. Subsequent chapters discuss simulations of the model that provide detailed quantitative analyses of linkages as well as income multiplers from variations in copper-sector variables on the rest of the economy.

Within the time period chosen for estimation of the model, the Chilean economy was subjected to sudden and extensive shocks resulting from basic changes in the political system. The most intensive changes occurred during the Allende period. This was followed by a period of countershocks, as the new government (1973-1976) attempted to reinstate the previous economic structure with emphasis on eliminating previous controls and regulations. In this context it was necessary to include in the specification of the behavioral relationships those variables which best represented the changes in the political system. Whenever these considerations were not feasible, dummy variables were used as proxies for those nonmeasurable or nonexisting variables.

Another important issue in the construction of any model for Chile is the persistent and erratically fluctuating inflation. For the period 1972-1975, it could be described as a hyperinflation. The underlying causes are complex, and much has been written on this subject with few consistent explanations. Expectations and other structural inflationary variables have been introduced in the relevant equations as an important element in the decision-making process of the economic units. In some cases, there is an essential consideration of structural lags in adjusting to unexpected inflation.

The description of the general structure of the Chilean model will cover some of the basic issues: the underlying theoretical assumptions, the sectors involved and their linkages, and the statistical methodology applied in the estimation. The existence of other econometric models of Chile for earlier periods (pre-1971 vintages) provide useful insights into the current work.[1] However, the basic objectives of this study and the time period of estimation differ substantially from those of earlier works. The following sections present some discussion of the assumptions and theory underlying the specification of the model. The second section of this chapter reviews the principal theoretical issues in constructing a model for Chile. The characteristics of the copper industry and the various linkages to the macro model are discussed in the third section. Another important feature of the model is the incorporation of supply constraints, and these are described in the fourth section. Finally, the last section presents the derivation of the behavioral equations of the multisectoral model.

Model Overview

The Chilean econometric model distinguishes eight sectors: (1) value added, (2) demand, (3) copper, (4) prices, (5) government, (6) foreign trade and balance of payments, (7) employment and factor income, and (8) money. The model is formed by 263 equations, of which 51 are behavioral and 212 are identities. The sample period for estimation of most behavioral equations is 1960–1976. All the equations have been estimated using ordinary least-squares method. The exogenous variables in the model include fiscal and monetary policy instruments, foreign-sector variables, and demographic trends.

Because of the importance of the copper sector for the present study, output and income determination in the copper industry's Gran Mineria have been disaggregated. Copper output has a direct impact on employment and value added in the economy. Inputs into copper production provide a link to utilities, transportation, and services value added. Other linkages are through copper-export revenues on the level of reserves and imports and through copper taxes on government investment, government purchases of goods and services, and transfers to individuals. The next section on the copper micro sector will provide more discussion on the role of copper in the Chilean economy.

An important feature in the construction of this model is the introduction in the model structure of a consistent set of production technology constraints. The employment, investment, overall price, and potential output equations are estimated with the help of a unique set of constant elasticity of substitution (CES) production-function parameters. The availability of factors of production, capital and labor, determines potential output for the economy. The ratio of actual to potential output gives a measure of capacity utilization, which enters the determination of the price equation and investment function. As capacity utilization is drawn to a maximum, prices are pressured upward, operating as a constraining force on the demand elements that determine actual output. The ceiling given by the potential output is eventually overcome by increased investment expenditures responding to the excess demand.

These adjustment mechanisms reflect the hypothesis that the long history of inflation in Chile may, in part, be the result of structural ridigities and persistent bottlenecks caused by the unavailability of capital and a lack of dynamism exhibited by the agricultural sector. Expansionary monetary policy has often accelerated the inflationary spiral by creating excess demand for goods, and because of the scarcity of factor inputs and the inability to finance more imports, the economy has had to respond by further increases in the rate of inflation.

The short-run determination of output is made through supply variables in some cases, but through demand variables for some of the most important sectors such as manufacturing and services. However, these demand-driven formulations do not prevent the existence of an upward-sloping short-term supply function for the economy, since (ceteris paribus) increases in actual output are set in motion by higher prices, which simultaneously increase capital formation and potential output, thus making possible further increases in actual output.

The production structure of the model describes the value-added components of the GDP. These are agriculture, Gran Mineria (copper), the rest of mining sector, manufacturing, construction, public utilities, and services. The specification of the agriculture, Gran Mineria, and public utilities sectors is made through supply variables. Manufacturing, construction, and services sectors are determined by demand elements.

The expenditure components of GDP are private- and public-sector consumption, gross fixed investment, inventory investment, and exports and imports of goods and services. Private-consumption expenditures on goods and services are determined by current and past income and inflationary expectations. The latter has been an important factor in view of the great uncertainty created by erratically fluctuating prices and abrupt social and political changes. Public-sector expenditures on goods and services consist primarily of wages and salaries to government employees.

The equation for gross fixed investment (this item is disaggregated into copper and noncopper investment) is an important element in determining the growth path of the economy. Because there are no consistent historical series on private or public investment, it was not possible to disaggregate this important expenditure item. The specification of the investment function incorporates explicitly the decision-making process of the government through the current account surplus/deficit of the public sector as an explanatory variable. The latter represents the availability of resources to be allocated to the capital account. In addition, the current account surplus/deficit determines the real flow of credit from the Central Bank to the public sector, thus providing additional financing of the investment program. The more current account savings that the government is able to generate, the less the need for financing through the banking system. Another aspect of investment that is influenced by decisions within the public sector is that pertaining to public firms, decentralized agencies, and public financial institutions. These enter the investment equation through the price of output and the rental price of capital terms. The rationale for distinguishing this element from the previous budgetary considerations is that the lack of domestic financing, economies of scale, or inadequate rates of return to compensate risks have deterred the growth of private investment. In order to offset the decline in the capital-output ratio of the private sector, the government has responded by expanding its investment into those areas which have traditionally been associated with private-sector investment in the more industrialized economies.[2] In this sense, a part of the underlying decision-making process for public investment would parallel that of the private sector in terms of maximizing expected future returns from an initial investment expenditure. The explanatory variables in the investment equation are desired capital (which depends on the ratio of the price of output and the rental price of capital), government savings in constant dollars, and capacity utilization. The government-savings variable establishes a link between public-sector budget and investment expenditures. Net investment measures the growth of capital stock, which, as a factor of production, determines actual and potential output. The capital stock also influences productivity and the real income of workers.

The balance of trade is a particularly important variable for Chile because imports represent an essential input into the production process. Imports are primarily determined by demand variables given import taxes and the exchange rate. In Chile, the possibilities for competing successfully with imports are limited. Import-substitution policies in effect for some time have created a need for other goods that had to be imported, such as raw materials. Imports of fuels are determined by domestic output requirements, while consumption and food imports respond to overall demand for goods. Most of the capital goods imports are dependent on the level of investment; the domestic market for capital goods would not justify the establishment of a diversity of capital-goods industries.

With the exclusion of copper, exports in general have not been a dynamic factor in the growth of the economy. One of the most important items has been wood and wood products. Chile has ample forest lands yielding good quality products that have been competitive in world markets. Noncopper exports are grouped into two categories: agriculture and others. The latter responds primarily to demand conditions in the industrial countries. A large proportion of the fluctuations in export earnings can be explained by the performance of traditional goods: copper, iron, nitrate, iodine, moylybdenum, fish meal, paper, and cellulose.

In the short run, the intersection of the aggregate demand and supply curves determines equilibrium output. However, the existence of supply constraints and the persistent disequilibrium in the money market prevents the economy from achieving equilibrium. Partial adjustments to changes in underlying variables, as represented by the lagged explanatory variables in the equations, cause the economy to move in the direction of equilibrium without attaining its goal. Because of tight regulations on interest rates that have maintained them at negative real rates, the money market rarely has achieved equilibrium. The increases in the money supply that originate from the additional financial requirements of the public sector have far outpaced demand for cash balances. This excess supply of monetary balances led to greater demand for goods and services, and the insufficient supply of goods led to significant increases in overall prices. Some of the excess demand is satisfied through increased imports of goods and services. These important linkages are represented in the model through the determination of the money supply and its inclusion, together with capacity utilization, in the behavioral equation determining the general price level.

Moreover, the external disequilibrium is linked to price determination. Whenever the disequilibrium in the goods and services market is transmitted to the balance of payments, there is a net increase or decrease in the level of international reserves. This variable determines the monetary base, so that external disequilibria are linked to price determination. The mechanism operates whenever there is an excess supply of money that is exchanged for more goods and services, thus exerting an upward pressure on prices. The interest rates have never responded to demand for nor supply of money. There is no market-clearing mechanism; instead, whenever supply of money is greater than demand, the excess balances lead are used to demand goods and services.

In addition to the effects of external inbalances on prices as the economy approaches full capacity utilization, the excess demand leads to an increase in prices through the overall price-level equation. Nominal wages are a function of productivity and inflation, and by determining the unit labor cost for producers, they are also an important explanatory variable in the price equation.

As exogenous variables, the international price of copper and import prices are important elements in generating cyclical behavior. The business cycles in the more industrialized economies are transmitted to the less developed via prices of important export commodities or of essential imports such as oil and other inputs. The level of import prices determines the demand for imports as well as overall domestic prices. The impact of fluctuations in the international price of copper on the rest of the economy can be traced through the copper sector and its effect on other sectors.

Micro Model of the Copper Industry

The copper sector of the model has been defined in terms of the so-called Gran Mineria del Cobre (for short, Gran Mineria), a term used in Chile to describe the operation of the large copper mines, which comprise around 83 percent of total copper output in the country. These mines have been subject to particular legal status and regulations, have had a majority of foreign ownership in the past, and are operations clearly profitable with low operating costs. Because of the importance of the Gran Mineria in total output and the existence of sufficient data on the larger mines, specification of the micro sector equations has been made in terms of the Gran Mineria. There are four different aspects to be considered in the copper sector: (1) production, (2) inputs (material, labor, and capital), (3) return to factors of production, and (4) government revenues. Between 1960 and 1976, various changes occurred that significantly affected the performance of the industry within each of these areas. In 1967, during the Frei administration, the Chilean government began to acquire a 51 percent ownership in the primarily U.S.-owned large-scale mines. A part of this program was an increase in investment expenditures to expand productive capacity. In 1972, the Allende government nationalized the Gran Mineria, and the public sector began to receive 100 percent of the profits, whereas previously a large percentage had been remitted to the U.S. owners.

Before presenting the mechanics of the copper micro sector, it is important to consider its role within the overall economic structure. The extent to which copper affects the other variables in the economy is governed by the importance of the two types of linkages: financial and real.[3] The financial variables link the copper sector to the public sector via revenues that are spent on goods and services and via balance of payments through export revenues, which are then used to increase imports, finance the external debt, or increase foreign-reserve holdings. The real effects pertain to employment and real value added by those sectors which provide inputs into copper production. The real linkages between copper and the economy include the following items:

1. Demand for physical inputs
2. Demand for labor and capital
3. Generation of wage and nonwage income
4. Demand for transportation services
5. Refining

Most of the ore that is not refined by the Gran Mineria is refined by ENAMI, a government enterprise that also does refining for a large proportion of the output from the medium- and small-scale mines. The financial linkages via the balance of payments (that is, those balance of payments variables which are determined by copper production and often affect the rest of the system) are

1. Imports of physical units and capital equipment
2. Refining abroad
3. Transportation costs
4. Interest payments
5. Remittance of dividends abroad
6. Export revenues

These variables comprise, in part, the main items of the balance of payments. Thus any changes in the copper sector will influence all the important components: the merchandise balance, nonfactor and factor services, the current-account balance, long-term capital account, and changes in the level of international reserves.

Finally, the linkages to the public sector are established through indirect taxes, direct taxes, and exchange-rate policies with respect to the Gran Mineria. The implied tax from the differential exchange rate is negligible during the 1960s. However, during the Allende administration and the first year of the Pinochet government, the Gran Mineria exchange rate was maintained significantly below the implied purchasing-power parity rate. As a result, the domestic cost component experienced an enormous increase.

Realized Export Price

Assuming that production decisions are guided by profit maximization, that is, marginal costs, the relevant output price is the realized export price. The latter is an endogenously determined variable whose value relative to the international reference price is influenced by the marketing strategies adopted by producers and the quality of Chilean ores. The realized export price is made a function of the LME cash price on which most contracts are based. The relationship between the two prices also may be affected by extraneous factors, such as the particular definition of the published LME price indicator and the monthly distribution of Chilean sales on the exchange. The LME price used in the model is a simple 12-month average, thus assuming a uniform distribution, whereas Chilean sales may be concentrated in certain months of the year, so that the realized price would not coincide with a simple average.

Before 1968, export prices for the Gran Mineria were a function of the U.S. producers' price. Afterwards they were tied to the LME price. In the case of medium- and small-scale mining, the export price has been mainly determined by the spot LME price. The estimated price equation for the Gran Mineria is

$$PECUGM\$ = 13.6228 + .657572PCULME\$ + .0236633PCULME\$(-1)$$
$$\quad\quad\quad (2.123) \quad (6.91) \quad\quad\quad\quad (.25)$$

$$- 10.3447DUM60/67$$
$$(-3.19)$$

$$\bar{R}^2 = .913$$

where PECUGM\$ is the realized export price for the Gran Mineria, and PCULME\$ is the LME spot price. The coefficients of the current and lagged LME prices were estimated using a second-degree Almon lag and constraining the right endpoint to zero. The dummy variable for the period 1960-1967 (DUM60/67) accounts for the setting of Chilean export prices according to the U.S. producers' prices. Since 1968, all transactions have been based on the LME spot price. The price equation for medium- and small-scale mines is

$$PECUPMM\$ = 3.92507 + .77708PCULME\$ + .033874PCULME\$(-1)$$
$$\quad\quad\quad\quad (1.89) \quad (16.92) \quad\quad\quad\quad (.765)$$

$$\bar{R}^2 = .971$$

where PECUPMM\$ is the realized export price for the medium- and small-scale mines. The value of the coefficients reflects the influence of freight costs, quality differential, and marketing strategy.

The output equation for the Gran Mineria follows, in part, the approach of Fisher, Cootner, and Baily (1972), where output responds to prices. The quantity of copper that producers want to supply will depend on some measure of profitability. This can be represented by the ratio of the realized export price to unit costs, which is a rough measure of the markup or the potential profits. Although Chilean exports have represented about 12 percent of total world market, their behavior is closer to that of a perfect competitor than to that of an oligopolist. Because of the significant lags involved in the materialization of targeted output levels and the large capital requirements, production decisions are dependent on the ability to predict future prices. This is incorporated in the supply equations through a formation of price-expectations mechanism.

Equations were estimated for the copper output of the Gran Mineria as well as for that of the medium- and small-scale mines. Production of copper in metric tons is a function of lagged relative prices, since planned changes in production take time to materialize. For the Gran Mineria, the relative price term is defined as the ratio of the realized export price to the cost of production, and for the medium- and small-scale mines as the ratio of the realized export price to the GDP deflator. In both cases, production was assumed to depend on an infinite weighted average of past prices, where the weights are assumed to correspond to a declining geometric series. The equation that showed the best results was linear in both the parameters and the variables:

$$CUPRODGM = \underset{(1.55)}{200.27} + \underset{(1.3)}{.63937}PCUGM/COST + \underset{(1.52)}{.37634}CUPRODGM(-1)$$

$$+ \underset{(2.79)}{10.254}TREND$$

$\bar{R}^2 = .717$
$D.W. = 1.64$

The supply price elasticities are .114 in the short run and .18 in the long run. The short-run elasticity is almost identical to the one obtained by Fisher, Cootner, and Baily; however, for the long run, their estimate is somewhat higher. The trend variable has been included to measure the impact of technological change on productivity.

The cost structure for the medium- and small-scale mines consists primarily of domestic inputs. Technology is, on average, rudimentary and highly labor-intensive. Given the large domestic component, the GDP price deflator is used as a proxy for the cost index. The estimated equation is

$$lnCUPRODPMM = \underset{(1.5)}{.53797} + \underset{(1.4)}{.140431}\, lnPPMM/PGDP$$

$$+ \underset{(12.1)}{.786698}\, lnCUPRODPMM(-1)$$

$\bar{R}^2 = .927$
$D.W. = 2.24$

The short-run price elasticity is .14; for the long-run it is .66.

Total costs of production for the Gran Mineria are defined as the sum of imported and domestic inputs plus the wage bill. Intermediate inputs into copper production, both imported and domestic, are calculated through input-output coefficients derived from historical production data.[4] Since the late 1950s, the government, through CODELCO, has sought to change the composition of inputs into the Gran Mineria in favor of the domestic component. Various incentives were offered to domestic producers in order to increase their participation as suppliers of the copper industry. The domestic-inputs coefficient serves as a policy instrument in the model; by varying the coefficient it is possible to measure the impact of these policies on the rest of the economy.

Employment

The employment equation for the Gran Mineria is a partial adjustment process between desired demand and actual employment:

$$\text{NPEGM} = \sum_{i=0}^{\infty} \delta(1-\delta)^i \text{NPEGM}^*(-i) + u$$

where NPEGM   = employment in Gran Mineria
       NPEGM* = desired demand for labor
       δ = partial adjustment coefficient
       u = random error term

The number of workers employed is some proportion of the difference between the number that the producers would like to hire based on their optimal output decisions and the number of workers on hand at the end of the previous period. The desired demand for labor NPEGM* is determined as

$$\text{NPEGM}^* = \text{CUPRODGM}/(Q/\text{LGM})$$

where CUPRODGM = copper production by Gran Mineria, in 000 M.T.
       Q/LGM   = output per worker of copper in the Gran Mineria

The denominator Q/LGM is the implied productivity of labor obtained from the labor marginal productivity condition under profit maximization behavior by the copper producers assuming a variable elasticity of substitution (VES) production technology. The VES production function differs from the CES in that it has a variable elasticity of substitution. The function can be written as

$$Q = \gamma \left[\delta K^{-\rho} + (1-\delta)K^{-m\rho}L^{\rho(m-1)} \right]^{-1/\rho}$$

where K   = capital
$\quad$ L   = labor
$\quad$ δ, ρ, m, γ = production-function parameters
$\qquad$ The elasticity of substitution is derived as

$$\sigma = 1/(1 + \rho - m\,\rho/\alpha_k)$$

where α_k is the elasticity of output with respect to capital.[5] The first-order condition for profit maximization yields an equation of the type

$$\ln(Q/L) = \text{constant} + \frac{1}{1+\rho}\ln\frac{w}{P} + \frac{m\rho}{1+\rho}\ln\frac{K}{L}$$

This formulation is somewhat restrictive through the assumption that social security and other payroll contributions by employers are not as important in calculating overall costs and hence are not included in the definition of wage per worker. The real wage term in the equation should be a measure of the real long-run expected wage rate. Incorporating a simple price-expectations mechanism, the regression estimation yields the following equation:

$$\ln(Q/LGM) = \underset{(-.3)}{-.27446} + \underset{(1.4)}{.07907}\ln W/PGM(-1) + \underset{(2.9)}{.33966}\ln K/LGM$$

$$\underset{(3.6)}{- .18711 DUM71/73}$$

$\bar{R}^2 = .721$
$D.W. = 1.99$

The dummy variable DUM71/73 measures the negative impact on productivity of the nationalization policies and the lack of a profit-maximization strategy of the Allende regime. Another factor in that period was the political turmoil that resulted in prolonged strikes interrupting the normal operating cycle.

$\qquad$ Performing a Koyck transformation on the employment partial adjustment equation with the infinite lag structure on NPEGM* yields the following form:

$$NPEGM = \delta NPEGM^* + (1 - \delta)NPEGM(-1)$$

The regression equation was estimated without the restriction that the coefficients sum to 1 in order to test the validity of the original specification. The estimated labor equation is

$$NPEGM = \underset{(.8)}{1080.8} + \underset{(1.5)}{.23995 NPEGM^*} + \underset{(3.9)}{.73786 NPEGM(-1)}$$

$\bar{R}^2 = .952$
$D.W. = .87$

The coefficients have the expected sign and the sum of the two is 1.021—quite close to the expected value of 1.

By assuming a VES production-function process, it is possible to measure the impact of an increase in the capital stock through investment on the supply of output equation through the costs of production. The accumulation of investment expenditures will increase the stock of capital, and if the capital/labor ratio increases, there will be an increase in productivity of labor and thus a decrease in unit labor costs. The lower unit costs will improve the profitability of production, as measured by the relative price term in the supply equation; this will increase the volume of copper production.

Investment

The decision to invest by the Gran Mineria depends on the expected present value of all future net revenues generated by the increased output. The criteria is based on a comparison of the present value of net of tax income generated by the investment with the initial required outlays; given a constant rate of discount, the net present value will vary with the discounted future price-to-cost ratio. The key decision variable affecting future revenues is the expected realized export price. Because of unpredictably wide fluctuations in the international price of copper, the success of investment programs has been determined principally by the ability to anticipate future price changes. The relative profitability variable, which determines investment expenditures, is the export price divided by an index of costs of capital goods used in the Gran Mineria. In order to account for labor installation costs for construction as well as machinery and transportation expenses in addition to the base price of producers' goods and materials, the latter is measured through the inflation index for the economy (the GDP price deflator). The resulting estimated equation for investment is

$$\ln \text{IBGMR} = \underset{(-3.64)}{-22.5746} + \underset{(4.24)}{3.70973} \ln \text{PCUGM/PGDP}(-1)$$

$$+ \underset{(.51)}{.626926} \ln \text{PCUGM/PGDP}(-3)$$

$$- \underset{(-2.32)}{1.69091} \text{DUM71/73}$$

$\bar{R}^2 = .694$

$D.W. = 2.87$

The lag structure chosen was the one that gave the best fit of the equation. From the size of the coefficients for the relative price term, it can be inferred that realized expenditures respond swiftly to changes in prices. During the Allende administration, imports of capital goods for the Gran Mineria were curtailed as a result of limited access to traditional suppliers' markets. Political conflict coupled with flight of personnel and inexperienced administrators delayed the implementation of the capital-budget program. These developments were accounted for in the equation through the introduction of a dummy variable DUM71/73 for the period 1971–1973.

An alternative formulation would be to consider copper investment as an exogenously determined variable. This approach could be particularly suitable to analyze the effect on the future economy of alternative copper investment programs.

Factor Income

The income side of the copper sector-accounts, or return to factors of production, is composed of the wage bill and other factor payments. These comprise the factor linkages of the micro sector to the rest of the economy. The wage bill is determined by the following identity:

$$WBILLGMN = (WGMN \times NPEGM)/1000$$

Nominal wages WGMN are negotiated through workers' unions in response to overall domestic inflationary pressures and copper-industry profits in an effort to secure the workers' share of the increased monetary benefits. Workers in the copper sector have traditionally earned significantly higher wages than workers in other sectors because of the much higher profits earned by that industry, strong labor unions, and strong bargaining power. The nominal wage equation is

$$\ln WGMN = \underset{(-28.1)}{-7.42232} + \underset{(17.4)}{.774158} \ln PGDP + \underset{(3.5)}{.16667} \ln OFPGMN71/72(-1)$$
$$+ \underset{(3.95)}{.51338} DUM71/72$$

$\bar{R}^2 = .996$
$D.W. = 1.41$

where   OFPGMN = nonwage income in the Gran Mineria

The dummy variable DUM71/72 represents the redistribution policies of the Allende government. During that period, workers' participation in industry profits was increased through higher wages.

The other component of factor income, nonwage income, is determined as a residual between the copper-sector value added and wage bill and capital consumption:

$$OFPGMN = X2GMN - CCAGMN - WBILLGMN$$

This variable serves as the tax base for the Gran Mineria. Before 1971, most of the other factor payments were designated as factor payments abroad, mainly to the Anaconda and Kennecott Companies, which owned the Gran Mineria complex. After 1971, nonwage income went directly to the government via income from public firms and properties. Indirect taxes are included in this aggregate nonwage income variable because of the lack of data on indirect taxes from the Gran Mineria.

During 1971 and 1972, the high wages and the exchange rate policies applied to the Gran Mineria led to increases in the cost of domestic inputs. In addition to higher costs, the operation of the mines was curtailed when Anaconda, whose mines had been nationalized, sought to prevent the sale of material inputs, spare parts, and machinery from the United States to Chile in response to the Allende administration's denial of compensation for its mines in Chile. In addition, higher operating costs resulting from inefficient administration and the lack of some essential materials coupled with low international copper prices led to a sharp decline in nonwage income in 1971 (about 95 percent from the previous year) with only a minor increase in 1972.

Government Revenues

Finally, the fourth main area within the copper micro sector is the determination of government revenues. Public-sector statistics on current revenues does not show the breakdown of copper revenues into indirect and direct taxes. The other link to government revenues is through the component designated as income from public firms and property. The value of these revenues has become relatively more important than the other sources of taxation since the nationalization of the Gran Mineria in 1971. The government revenue equations estimated for the model are direct and indirect taxes, while income from public firms, which is largely dependent on government policies, is considered an exogenous variable.

Because direct and indirect taxes are not divided into copper and noncopper, these totals were made a function of the copper and noncopper components. Direct corporate taxes are a function of the tax base defined as nonwage income. The functional relationship is specified as

$$TXDCORPN = f[OFPN, OFPN(-1)]$$

Indirect taxes are determined through an identity:

$$TXINDN = TXRATEIND \times X123NTN$$

where TXRATEIND is the implicit tax rate assumed to be an exogenous policy variable, and X123NTN is the sum of the nominal value added for primary, secondary, and tertiary sectors. These formulations assign an implicit role to the Gran Mineria in the determination of the overall level of taxes. The explanatory variable in the direct tax equation OFPN is the sum of the nonwage income for the Gran Mineria and for the rest of the economy.

Copper Export Instability

The specification of the behavioral equations for the copper micro sector has incorporated most of the relevant linkages, both real and financial. Another possibly relevant consideration is the relationship between domestic activity and export instability. Copper accounts for about 60 to 70 percent of total exports, so that persistent pronounced fluctuations would have an impact on the rest of the economy. Instability can perhaps be described as large unanticipated fluctuations in earnings. Uncontrollable variations in the price of a commodity are not necessarily detrimental to the growth of an economy. Just as an individual can implement protective measures to safeguard an investment by diversification of risky assets, an economy can offset the potentially negative effects of instability through prudent fiscal and monetary policies. This section provides a brief description of a test of the significance of export instability.

A survey of the literature by Manger (1979) outlines some potential linkages from export instability. The variables that are more likely to be affected are investment, government expenditures, and imports of goods. The uncertainty about the economic environment or the availability of capital that arises from export instability could deter investment expenditures. The public sector also might reduce expenditures on current account whenever greater export instability became apparent, if it is politically feasible. Such a measure could prevent sudden cutbacks in public investment or use of inflationary financing. Finally, the impact on imports could vary depending on the type of good. Imports of materials could increase with instability if producers acted to offset the risks (if they exist) of a reduction in the supply of foreign inputs. Consumption-goods imports could be negatively correlated with instability if consumers ascribed domestic instability to export instability.

In order to test whether export instability has been a relevant factor in

Chile, the following measure of instability was included in the behavioral equations for each of the preceding cases:

$$\text{Variability} = \left\{ \left[\sum_{i=1}^{5} (X_{t-i} - \bar{X}_t)^2 \right]^{1/2} \middle/ 5 \right\} \middle/ \bar{X}_t$$

where X_t = variable whose "variability" is measured by the formula

$$\bar{X}_t = \text{5-year moving average } (X_{t-1} + \cdots + X_{t-5})/5$$

There were two definitions of X_t used: one was the current U.S. dollar value of total exports, and the other was the current U.S. dollar value of copper exports. Another variable that would incorporate some of the export instability is the level of international reserves. The preceding formula is equivalent to the sample standard deviation divided by the mean of the variable.

Table 3-1 shows the results of the estimation of five behavioral equations with respect to the variability index coefficient. The five equations are copper and noncopper gross fixed investment, government consumption, imports of consumption goods, and other material imports, where each variable is measured in real 1965 pesos. A discussion of the theoretical specification for each equation is presented in the following section. Out of fifteen trials, three have statistically significant coefficients, but in all three cases, the sign was "wrong." In none of the investment equations is the measure of variability significant. The variability of international reserves is significant with respect to government consumption, but it has the wrong sign. Copper-exports variability is significant with respect to imports of consumption goods, but it also has the wrong sign. The inclusion of either of the two export-variability measures changed the sign of total

Table 3-1
Chile, Results of Estimated Equations with Variability Term

	$EXVAR^a$	$CVEXVAR^b$	$RESVAR^c$
Noncopper investment	i(−)	i(−)	i(+)
Copper investment	i(+)	i(+)	i(+)
Government consumption	i(−)	i(−)	s(+)
Imports of consumption goods	i(+)	s(+)	i(−)
Other material imports	i(+)	i(+)	s(−)

Note: For each behavioral equation, the coefficient is shown as significant s or not significant at the 10 percent level of significance. The sign of the coefficient is in the parentheses.
[a]Variability of total US$ exports.
[b]Variability of total US$ exports.
[c]Variability of US$ reserves.

consumption expenditures from positive to negative. Finally, the only significant result for other material imports was obtained with international reserves with a negative, instead of the expected positive, sign.

Of the three cases with statistical significance but with the wrong sign, two are represented by the measure of international-reserves variability. This is not a valid measure of export instability. This latter variable refers only to unanticipated fluctuations in the price or quantity of the commodity export. The variability of international reserves embraces both export and political instability. In the case of Chile, the dramatic political changes have affected the availability of international reserves. The success of the monetary authorities in stabilizing domestic activity and prices during the three political periods is another element altering the level of reserves independently of the world copper price and Chilean production.

Structure of the Model

Supply Constraints

The global production function for the noncopper sectors represents the collection of all efficient combinations of output for a given set of inputs. It is a mathematical function relating quantities of inputs, or factors, to output. The inverse of this function represents the input requirements necessary to produce a given quantity of output. By estimating an aggregate production function for Chile, it is possible to integrate the market for factors of production with the goods and services market. Through this linkage the occurrence of any bottlenecks on the availability of factors of production imposes a constraint on the domestic output of goods and services, which induces higher prices and more imports. The aggregate production function for Chile describes the technological constraints as well as the potential growth path in response to the existing technology.

The mathematical form of the production function is that of the constant elasticity of substitution (CES) production function. This function is specified as:

$$Q = \gamma(\delta K^{-\rho} + (1 - \delta) L^{-\rho})^{-\nu/\rho}$$

where γ = scalar, denotes efficiency
δ = distribution parameter, represents degree of capital intensiveness
ρ = substitution parameter
ν = degree of homogeneity, or returns to scale
K = utilized stock of capital
L = labor input

The distinction between the CES and VES production functions for the aggregate and the Gran Mineria output is that technology in the latter sector is much more homogenous, permitting the use of the more sophisticated function.

The elasticity of substitution between capital and labor is derived from the substitution parameter by the following expression

$$\sigma = \frac{1}{1 + \rho}$$

Because of the number of parameters to be estimated, it is not possible to identify each one from the preceding production condition. It is necessary to introduce several a priori restrictions that would result in the identification of the production function and its parameters. The underlying economic assumption is profit-maximization behavior. By assuming constant returns to scale for aggregate output in Chile ($\nu = 1$), the marginal-productivity condition from the profit-maximization strategy becomes

$$(1 - \delta)\gamma^{-\rho}\left(\frac{Q}{L}\right)^{1+\rho} = \frac{W}{P}$$

By taking logarithms of the latter expression and simplifying,

$$\ln(Q/L) = \sigma[\rho\ln\nu - \ln(1 - \delta)] + \sigma\ln(W/P)$$

An estimate of this equation would yield a value for the elasticity of substitution parameter. The relationship estimated for Chile is based on noncopper aggregate value added:

$$\ln \text{GDPNGM/NPE} = .73771 + .17485\ln \text{W/PGDPNGM}$$
$$(5.3) \qquad (3.2)$$

$$+ .51497 \ln \text{GDPNGM/NPE}(-1)$$
$$(5.4)$$

$\bar{R}^2 = .942$
$D.W. = 2.53$

The inclusion of the lagged dependent variable results from differentiating between the short-run and long-run effects. The implicit assumption is that the possibility for substituting capital for labor is more restricted in the short than in the long run. This is a more realistic assumption in view of the relative heterogeneity of factors of production and the learning time required in adapting them to new production activities. The time period for estimating the production function is 1960–1971. The exclusion of the Allende years in the estimated equation avoids the possibility of distortions from the extensive redistribution of income in favor of wage earners during the 1971–1973 period. This redistribution did not create a fundamental change in the structure of production in the economy. During that period, the economy operated close to the production frontier described by the same production-function parameters as in the previous two decades. The estimated value for the elasticity of substitution between capital and labor in the short run is .17485; in the long-run it is .36049. From time-series data on factor shares, it is possible to derive the capital intensity, or distributional, parameter δ. The sample estimate was obtained by the arithmetic mean of the ratio of other factor payments to the sum of other factor payments and salaries. So $\hat{\delta}$ = .55, and the corresponding labor intensity, or distributional, coefficient is .45. After the application of the previous two independent production conditions, it is possible to estimate the efficiency parameter to complete the specification of the CES production function. This step leads to the following regression equation:

$$GDPNGMR1 = .577M$$

$$\bar{R}^2 = .95$$
$$D.W. = .98$$

where $M = (.55 \times KUTNGM^{-1.7695} + .45 \times NPENGM^{-1.7695})^{-.5651}$

Since the production function describes the amount of factor inputs used in producing a given quantity of output, the capital-stock series had to be adjusted to represent the proportion actually utilized in production. By assuming a fixed proportional relation between capital and labor, the stock of utilized capital was defined as the existing stock minus a percentage of the total, given by the rate of unemployment.

The final form of the production function for non-Gran Mineria GDP is

$$GDPNGMR1 = .2577 \times (.55 \times KUTNGMR^{-1.7695}$$
$$+ .45 \times NPENGM^{1.7695})^{-.5651}$$

The ratio of actual to potential output is the rate of capacity utilization, which is an important mechanism for adjusting output whenever excess demand or supply exists. The need for this variable arises from the conflict between the demand-determined output model of the Keynesian tradition and the chronic supply constraints in developing economies despite an overabundance of unskilled and semiskilled labor.

Actual output in the model is determined through the value-added functions. As explained previously, the sectors included are agriculture, Gran Mineria mining, non-Gran Mineria mining, construction, manufacturing, public utilities, and services. These, in turn, are dependent on demand and supply variables. If the rate of capacity utilization increases significantly above the period average, investment and the overall price level respond; the latter acts as a ceiling, whereas an increase in the former expands the productive capacity of the economy with a lag and reduces future rates of capacity utilization.

In the long run, growth of output in the economy is a function of the growth rates of capital stock and the labor force. This condition is fulfilled through the capacity-utilization variable. While some components of output are determined by demand in the short or even medium horizon, in the long-run equilibrium path, the demand and supply components must grow at equal rates. Any cyclical deviation of demand for output from potential supply will disturb the equilibrium rate of capacity utilization and initiate a gradual adjustment of the rate of growth toward the long-run equilibrium rate.

Aggregate Demand

This section describes the structure of the aggregate demand function of the economy, which has been disaggregated into private- and public-consumption expenditures, gross investment (non-Gran Mineria), inventory investment, and imports and exports of goods. The specification of public-consumption expenditures is discussed in the section dealing with the public sector and the import and export functions are discussed in the section dealing with the foreign sector.

Private Consumption. In the consumption function, the dependent variable is defined as real private consumption expenditures on goods and services. The measure of disposable income used in the estimation was divided into two components: wage and nonwage income.[6] In the case of Chile, this distinction has some particular advantages. One follows from Marx's and Kaldor's hypothesis that workers consume most of their income, while capitalists save a large proportion. Thus there is the possibility of two groups with significantly different consumption behavior. In view of the historical events in 1971 to 1973, where a marked shift occurred from nonwage to wage income, this distinction is instrumental in explaining behavior in the real world. Another advantage is that by identifying the two classes of consumers, it is possible to evaluate the impacts of shifts in the distribution of income, an issue that is of a particular importance in a developing economy.

Friedman's permanent income theory provides the basis for construction of the income variable.[7] An individual's decision on how much to consume is based on a notion of permanent income. Transitory income, for example, a bonus, winning the lottery, or accepting a payoff, does not influence permanent consumption. The crucial term is *permanent.* In arriving at a notion of permanent income, an individual must evaluate past incomes and his expectations of the future earning stream. The bridge between theory and reality is built from past history. An average of x number of years is used as a formula for generating an individual's expectation of permanent income. A variety of alternative definitions is obtained by specifying different series of weights for different periods. Usually the formula is constructed by use of a declining set of weights reflecting the importance of the immediate past in providing information about the future. By assuming an infinite lag structure in the formulation of permanent income, where the weights are given by a geometric series, a simple and reasonable approximation to reality is obtained with a minimum of coefficients in the consumption function.

The permanent income definition based on the weighted sum of infinite past incomes reduces the number of explanatory variables in the consumption function to a measure of current income and lagged consumption:

$$C = f(Y,C_{-1})$$

This function describes how income determines consumption. Since the consumption theory deals with a single economic unit, the macro function represents the aggregation of all these units. Under suitable assumptions, this leads to a relationship in terms of per capita values.[8]

This basic behavior pattern has been modified in the Chilean experience as a result of the long inflationary history. Individuals have learned to take protective measures against the sudden erosion of their purchasing power. Since the financial institutions have not been able to provide adequate security for their real financial assets, the alternative has been to purchase goods as a hedge against inflation. The higher the inflation rate, the more goods individuals will purchase in anticipation of even more inflation. The lagged inflation variable represents a simple price-expectations formula that has been included in the equation to take this particular effect into account:

$$CEPR/N = .14 + .73WGSTOTNTR/N + .56OFPNNTR/N$$
$$(2.45)(9.56) \qquad (5.05)$$
$$+ .32CEPR/N(-1) + .00015PGDP^*(-1)$$
$$(3.94) \qquad (2.5)$$

$\bar{R}^2 = .984$
$D.W. = 1.912$

where CEPR/N = real per capita private-consumption expenditures
WGSTOTNTR/N = real per capita net of income tax wages and salaries
OFPNNTR/N = real per capita net of corporate tax nonwage income
PGDP* = overall rate of inflation

The short-run marginal propensity to consume out of wage income is .734, and for nonwage income it is .56. These are consistent with the hypothesis that individuals consume a greater proportion of wage than nonwage income. In the long run, assuming identical lag structures for both types of income, the marginal propensity to consume (MPC) of wage income is 1.08, and for nonwage income it is .82. The mean lag is half a year. A weighted sum of the long-run MPCs, where the weights represent the distribution of the mean values for the per capita income variables, yields an estimate of .943.

Private Investment. The specification of the investment equations is based mainly on the neoclassical theory of investment, in which investment decisions are explained as the process of adjustment to a desired level of capital stock (K*). In this neoclassical framework, the level of desired capital depends on the relative price of capital with respect to the overall price level and the value of output.[9]

Once the firm has decided to augment its existing stock of capital, the necessary purchases of capital goods must be approved as part of the capital budget, then the orders for the equipment are placed on suppliers' back log, and finally the items are delivered and payment is disbursed. The first, in effect, adjusts gradually to its target capital stock, where the rate of adjustment depends on the cyclical path of the economy. This relationship can be described as follows:

$$I_t = \sum_{j=0}^{\infty} \mu_j \Delta K^* = \mu(L)\Delta K^*$$

In the case of a CES production function (discussed in a previous section), the first-order condition can be written as

$$\frac{\delta Q}{\delta K} = \delta \gamma^{-\rho} \left(\frac{Q}{K}\right)^{1+\rho} = \frac{C}{P}$$

This expression can be solved for K, so

$$K^* = (\delta \gamma^{-\rho})^{\sigma} Q \left(\frac{P}{C}\right)^{\sigma}$$

where Q = gross output
       K = stock of capital
       C = rental price of capital
       P = price of output
       K* = desired K

To estimate the rental price of capital services for Chile, the following formula was applied:

$$C = PIB\left(\frac{\delta + r}{1 - TXRATE}\right)$$

where PIB   = investment deflator (national accounts)
       r = cost of capital
       TXRATE = direct corporate tax rate
       δ = depreciation rate

In the absence of an original market for capital services, the preceding formula is an attempt to approximate the implicit cost per time period of owning and operating capital equipment. The investment process in Chile has come to rely to a large degree on foreign financing. The ceilings imposed on domestic interest rates coupled with high rates of inflation have resulted in negative real rates of interest throughout the sample period. In order to offset the lack of savings, the government obtained credit from external sources or provided guarantees for private-sector borrowing abroad. These funds were then channeled through development banks to finance investment projects with a major external component. Under these circumstances, the cost of foreign capital could be a very relevant measure of the overall cost of capital in Chile. The long-term U.S. government bond rate is used as the interest rate on external financing. An alternative cost measure, the implicit rate from the balance-of-payments interest payments on medium- and long-term debt was not used because of the difficulty in separating out debt rescheduling and unreported capital transactions.

Finally, the investment function includes the rate of capacity utilization, since the level of desired capital stock is determined, in part, by the existing rate of capacity utilization as a measure of excess demand. The estimated non-Gran Mineria gross investment equation is

$$IBNGMR = \underset{(.32)}{.1586} + \underset{(2.07)}{.0428DK*(-2)} + \underset{(5.16)}{.8920IBNGMR(-1)}$$

$$+ \underset{(2.21)}{.0318DGDPNGMCAPUT} + \underset{(.83)}{.0557GVSVR(-1)}$$

$\bar{R}^2 = .744$
$D.W. = 2.01$

where DK*   = change in desired capital stock
       DGDPNGMCAPUT = non-Gran Mineria rate of capatity utilization
       GVSVR   = real current-account savings of the public sector

The equation was estimated using the Jorgenson lag structure that provides the link between the determination of desired capital stock and realized investment expenditures.

The capital stock is assumed to be of the putty-putty type.[10] Thus the estimation of the investment equation does not distinguish between replacement and net investment.

The lack of data on public-sector investment expenditures did not allow the estimation of separate investment functions for the private and public sectors. By attempting to provide incentives to investment and growth, the public sector has become a major savings and investing unit, even in areas that have traditionally belonged to the private sector in the more industrialized economies. The inclusion of the real level of public-sector savings on current account in the equation provides an essential linkage to government investment expenditures.

Inventory Investment. Investment in inventories is defined as the change in stocks of materials, goods in process, and finished goods. Because there are no data on each of these items, the estimated equation includes all three components. It is assumed that adjustments in the three levels of inventories are derived from the expectations about future sales. The general form of the stock-adjustment model is

$$ICHR = \alpha(INV^* - INV_{-1})$$

where ICHR = real inventory investment
         INV* = desired inventories stocks

The variable INV* represents the target level of inventories. After setting a target, the firm adjusts partially to that level, where the rate of adjustment depends on the production technology, availability of products, cost of capital, and storage costs. Since inventories eventually become part of sales by the firm, the decision on the desired stock arises from expectations about future sales.[11] A simple representation of this mechanism would be

$$INV^* = f(S,DS)$$

where S   = sales (GDP minus change in inventories.)
         DS = change in sales

The estimated equation is

$$ICHR = .73187 + .1906S - .05789DS - .481776INVR(-1)$$
$$\quad\;\; (1.0) \quad\;\; (1.4) \quad\;\; (-.6) \quad\quad (-1.3)$$

$$- .00057PGDP^*(-1)$$
$$(-1.0)$$

$\bar{R}^2 = .503$
$D.W. = 1.53$

Risk is also an important consideration in decisions concerning inventory accumulation. It can be associated with expected price inflation. Higher inflation would imply higher business risks and, hence, lower inventory accumulation. The risks associated with market conditions would lower demand for inventories of materials and intermediate goods as producers avert potential losses from higher storage costs as well as unstable demand for their products. However, it might be better to hold inventories than monetary assets. Taking a simple price-expectations formula where inflationary expectation for the current year is a function of the previous year's inflation, the resulting coefficient for the lagged inflation rate shows the expected negative sign in the inventory-investment equation.

Production Sector

The production of goods and services is divided into three sectors. In less developed countries, much effort has been devoted to the development of a dynamic industrial sector that would provide employment for a rapidly expanding labor force, supply inputs to and process the output from agriculture to help feed the population, and ease the balance-of-payments deficit through import substitution. Because of the strategic role of this sector, the value-added equations have been estimated for its components: construction, non-Gran Mineria mining, manufacturing, and public utilities. Agriculture and services equations are estimated separately. The latter includes some critical services, such as transportation, as well as a miscellaneous category of services, activity that absorbs the excess supply of labor unable to find work in the urban centers.

Agricultural Value Added. Primary-sector value added is determined through supply considerations. For this sector, capital may be the constraining factor of production. The excess supply of labor has resulted in the migration from rural areas to urban centers. The scarcity of capital and suitable land for the available farmers has deterred the growth in output needed to feed the growing urban centers. Another cause of displacement of agricultural output has been the relatively higher prices earned in the industrial sector, in contrast to the prices of farm products, which, in many instances, are subject to government regulations.

One of the explanations repeated in the migration literature is the unfavorable terms of trade for agriculture with respect to manufacturing. Farmers, disenchanted with their meager incomes, seek the higher standard of living in the cities. The large-scale agricultural producers, however, are quite responsive to the terms of trade in deciding whether to expand their operations or divert their earnings into other sectors. Since land reform has been a relatively recent phenomenon in Chile, there have been no significant structural shifts in the ownership and exploitation of resources until the latter part of the Frei administration and the subsequent large-scale expropriations during the Allende period.

The equation for primary-sector value added represents a hybrid of a supply and a production function. The specification is based on some notion of the structuralists' theories. The aggregate function is assumed to represent a typical supplier who controls the production of two outputs and two inputs. These correspond to agricultural and industrial production, given some allocation of capital and labor between the two. The production process is described as

$$F(X1,X2,K,L) = 0$$

where $X1$ represents primary output, and $X2$ represents industrial output. If the individual suppliers maximize their profit for given output and input prices, then the first-order condition can be written as

$$\frac{P1}{P2} = \frac{\delta F/\delta X1}{\delta F/\delta X2} = \frac{-\delta X2}{\delta X1} = MRTS$$

where MRTS is the marginal rate of technical substitution between primary and secondary output. As the prices of primary-sector goods increase relative to those for secondary goods, producers will substitute production of primary goods until the first-order conditions are reestablished. The preceding expression can be solved explicitly for primary-sector output as a function of relative prices once the production process is known. The supply response is thus represented by the relative price term.

The stock of capital can be reasonably assumed as the constraining factor in the production function. Given these assumptions, the specification of the value-added equation includes a relative price term and a capacity variable, defined as the stock of capital for non-Gran Mineria, representing the supply constraint. Since output does not adjust instantaneously to changes in prices and factors of production, it is assumed that actual output approaches the desired level following a geometric lag. The equation can be simplified through a Koyck transformation to yield the following estimate:

$$X1AGR = -1.4305 + .00565PX1AG/PX2MFG(-1) + .03153KNGMR(-1)$$
$$(-1.88) \quad (1.74) \quad\quad\quad (3.4)$$

$$+ .64154X1AGR(-1) - .3573DUM72/73$$
$$(2.7) \quad\quad\quad (-3.7)$$

$\bar{R}^2 = .778$
$D.W. = 2.4$

The short-run price elasticity is .241, the long-run is .672. As suggested by the structuralist theory, the elasticity for the long run is significantly less than 1, but it is not a very low value. This suggests some degree of response to profit incentives by the primary sector. The trend effect of population growth on agricultural output is incorporated via the equation for the primary-sector price deflator. One of the explanatory variables of this equation is the deviation of primary-sector output form its long-run trend. As growth of the population narrows the deviation, primary-sector prices rise, bringing about an increase in output.

The dummy variable in the estimated equation reflects the political turmoil surrounding the land-reform program. In contrast to the reform program of the Frei administration, during the Allende period, numerous revolts by the farmers resulted in large-scale expropriation of farmland. Large operating farms were divided up into many parcels that were then hastily distributed to farmers who lacked the appropriate tools and machinery for efficient exploitation of the land. The lack of organization for redistributing the land and the constant political squabbles resulted in significant reductions in actual output.[12]

Manufacturing Value Added. Production of manufacturing value added is a function of aggregate expenditures: consumption and investment. Manufacturing exports also can be included, but their volume and dynamism were negligible during the sample period.

This functional relationship must be thought of as a reduced-form equation of an input-output system in which value added can be expressed as a transformation of input-output coefficients and the final demand vector by assuming that value added is proportional to gross output. This approach is applied to manufacturing and services sectors, although in differing forms. The basic input-output relation is described here, and its application to the service sector is described in a following section. The relation between final demand and value added is written as[13]

$$X = \beta(I - A)^{-1}D$$

where X = value-added vector
   A = input-output coefficient matrix
   β = matrix (diagonal) of markup factor of value added to gross output
   D = final demand vector

This method of applying fixed input-output coefficients does not allow the possibility for changes in the pattern of intersectoral flows that occur through variations in relative prices in response to the supply constraints. However, in the short run, the input-output approach to determining value added is reasonable approximation to reality.

In the long run, output levels should depend on the supply of factors of production, capital and labor, and the particular composition of final demand. The value-added equation estimated for this model does not incorporate these long-run considerations explicitly, but the model as a system takes into account these variables through the notion of capacity utilization.

The positive price effect of capacity utilization is a response to supply limitations. In addition, its positive effect on investment alters the composition and level of final demand in order to generate sufficient productive capacity to meet future demands. These linkages between demand and output via the value-added equation are adequate for analysis of long-run growth paths for an economy with fixed coefficients in the short run.

The final demand variables in the value-added equation are total consumption expenditures and total gross domestic investment. The estimated equation is

$$\ln X2MFGR = \underset{(-6.5)}{-1.0629} + \underset{(10.8)}{.7475 \ln CER} + \underset{(5.7)}{.4729 \ln IR}$$

$$+ \underset{(1.6)}{.0458 DUM71/73} - \underset{(-2.7)}{.1196 DUM75}$$

$$\bar{R}^2 = .97$$
$$D.W. = 1.174$$

Because of the dynamic growth of this subsector, the underlying production relation is specified in log linear form. During the period 1960–1970, manufacturing value added grew at an average annual rate of 5.4 percent, while GDP grew at 4.5 percent. The coefficients of the estimated equation are related to the input-output table. Each one is a function of the corresponding input-output coefficient relating gross output to final demand.

The expenditure items included in the equation consist of a domestic and an imported component. The proper specification should include only the domestic component, but this is precluded by the lack of disaggregated information. Some bias should be expected in the estimation of the regression coefficients if the share between these two components varies significantly in the sample period.

It is possible to identify at least two periods in which this has taken place. The first one in the Allende period, where the sudden jump in consumption expenditures resulting from the sharp income-redistribution policies pursued was not accompanied by a proportional change in the imported components, hence implying a more than proportional response in the industrial output than what the increase in the total consumption expenditure figure would indicate. This effect was taken into account in the regression with the inclusion of the dummy variable DUM71/73, which took a positive sign, as expected. A completely opposite situation arose in 1975 when the country initiated the policy of trade liberalization. Under those circumstances, the composition of the expenditure items changed toward increasing the participation of imported goods, thus decreasing the demand pressure on the domestic manufacturing sector. Additionally, the trade-liberalization policies had the effect of eliminating the inefficient producers from the domestic sources of production that depended on protection from foreign competition. These developments were reflected in the estimated equation by the inclusion of the dummy variable DUM75, which took a negative sign.

Construction Value Added. By definition, value added for construction is equal to gross domestic investment in residential and nonresidential construction minus purchase of material inputs. However, the investment equations in the model correspond to the total for Gran Mineria and non-Gran Mineria. Because of data limitations, it was necessary to estimate a linkage equation between construction value added and gross domestic investment minus imports of capital goods. This residual is a very close approximation to fluctuations in construction expenditures, since imports of machinery account for a large percentage of investment in machinery and equipment. The estimated equation is presented in Appendix A.

Non-Gran Mineria Mining Value Added. The non-Gran Mineria mining sector includes copper production for the medium- and small-scale mines, iron ore, nitrates, and molybdemum. Gross output has been specified as a function of a capacity variable represented by the capital stock of non-Gran Mineria. The lagged dependent variable is included through the specification of a partial adjustment mechanism between expansion of productive capacity and actual output. The estimated equation is

$$X2MINNGMR = -1.69768 + .0431KNGMR(-1) + .4026X2MINNGMR(-1)$$
$$(-2.20) \quad (2.32) \quad (1.57)$$

$\bar{R}^2 = .931$
$D.W. = 1.59$

Public Utilities Value Added. Public utilities represents a very small component of total output, with an average share of about 2 percent between 1960 and 1967. Even though it represents a small share of output, it is an essential input into production. Value added by this sector is a function of demand considerations. This assumes that creation of generating capacity does respond to industrial demand, but as the equation shows, output by this sector adjusts with some lag. The explanatory variables are total value added by the primary and secondary sectors, excluding the Gran Mineria and public utilities, and production of copper by the Gran Mineria. The latter is a linkage from the copper micro sector to the macro model. From the data for 1976, energy inputs into the Gran Mineria accounted for about a third of its domestic costs.

Whenever demand expands too rapidly, the available supply is exhausted. In the very short run, with a given supply of output, the firms are unable to produce more in order to increase their profits. Electric-power generation is characterized by these constraints. Because of the time required for installment of new generators, producers adjust slowly to an increase in demand. The specification of the equation includes a partial adjustment process from desired to actual output:

$$X2UTR = -.14184 + .00025CUPRODGM + .01407X12NGMUTR$$
$$(-2.79) \quad (2.6) \qquad\qquad\qquad (2.8)$$
$$+ .68873X2UTR(-1)$$
$$(6.1)$$

$\bar{R}^2 = .973$
$D.W. = 2.2$

where CUPRODGM   = copper production by the Gran Mineria
          X12NGMUTR = primary- and secondary-sector value added, excluding
                                  Gran Mineria and public utilities

The elasticity with respect to primary and secondary sectors is .396 in the short run and 1.27 in the long run. For copper production, they are .40 and 1.28, respectively.

Services Sector Value Added. The tertiary sector value added is determined by the same input-output process as in manufacturing. That is, the final demand components determine output from services through the input-output coefficients. Additional variables influencing the level of services output are the level of foreign trade, imports plus exports of goods and services, through transportation insurance and storage requirements; and copper production through the transportation, retailing, banking and other services which make up purchases of domestic inputs and output. The inclusion of production by the Gran Mineria establishes a link between the micro sector and the macro model. The estimated equation is

$$X3R = .38063 + .32012C+I + .461325E+MGSR + .001224CUPRODGMAV2$$
$$(.99) \quad (12.7) \qquad (4.46) \qquad\qquad (1.3)$$

$\bar{R}^2 = .993$
$D.W. = 1.84$

The elasticity with respect to final demand is .63, and for external trade it is .262. These results are consistent with the structure of output of services that are dominated by the wholesale and retail businesses and miscellaneous services. They are primarily oriented toward the final goods markets. The enclave-economy phenomenon of a primary commodity exporting country is illustrated in part by the low value of the coefficient for copper production. The output elasticity with respect to this variable is only .07, which reflects the limited potential for expanding noncopper employment through the linkage between production activity in the Gran Mineria and the provision of domestic inputs. Whereas the value added of the services sector corresponding to copper production is a small proportion of the total, there are other linkages, such as copper exports (represented by the E+MGSR variable) and gross fixed investment in the copper sector (included in the C+I term), that have significant impacts on the rest of the macro model.

Foreign Sector

One of the potential constraints to economic growth in Chile is the balance of payments. The economy is dependent on imports for essential inputs into production and on other goods not available in the domestic market. Hence the capacity to import, as measured by the export earnings, plays an important role in determining the potential long-run growth path of the economy. The published data on imports include the following end-use categories: imports of food, consumption (excluding food), capital, and intermediate inputs. The latter are subdivided into petroleum and other fuels, materials for the Gran Mineria, and other intermediates. An import-demand equation has been estimated for each of these categories. This level of disaggregation permits a more sensible analysis of government policies through their impact on the underlying economic decision variables associated with each of the import-demand equations. Exports of goods are divided into agriculture, copper, and other. Throughout the period 1960–1976, copper represented more than 80 percent of total export earnings.

In general terms, an import-demand equation can be specified as

$$M = M\left(\frac{P_m}{P_y}, \frac{Y}{P_y}\right)$$

where M is quantity of imports; P_m is the price of imports; P_y is the price level of domestic goods, and Y is domestic money income.

Other variables not considered in the static micro-unit optimization problem are relevant at the macro level. Since the exchange rate is strictly controlled by the Central Bank, the "ability to import" concept is particularly relevant. The shortage of foreign exchange has forced the government to restrict imports through import tariffs, quotas, prior import deposits, and other schemes, creating a complicated network of legal regulations that is very difficult to implement. Because of the large number of policy variables applied to restrict imports and the lack of adequate data on the important ones, the level of international reserves is used as a proxy variable for the "ability to import."

The Allende and post-Allende periods brought drastic political changes

affecting fundamental policy variables. The pre-Allende period had been characterized by strong protectionist policies in an effort to promote the growth of
import-substitution industries. The widespread nationalization during the
Allende government, switching from private to government-owned industries,
resulted in bottlenecks that were intensified because of an insufficient supply of
essential imported inputs.[14] Finally, the Pinochet administration initiated a set
of new foreign-sector policies drastically reducing import tariffs and eliminating
inefficient controls. The effects of these political changes have been incorporated
into some of the import equations by specifying dummy variables during the
unusual periods.

Food Imports. One of the reasons for separating food from the other consumer
goods is the strong exogenous element in domestic food production and the
need to offset any decline in output during a bad crop year with an adequate
supply of imports. The impact of a bad or bumper-crop year on food imports
has been introduced in the equation through a measure of the deviation of agricultural value added over its long-run trend. The long-run trend agricultural
value-added variable is determined by an average exponential growth rate, which
is obtained from regressing agricultural value added on a time-trend variable:

$$X11GRLT = 1.64903e^{.017t}$$

The average long-run growth rate has been 1.7 percent per year.

For both food and nonfood import equations, total real consumption expenditures has been used as the activity variable instead of income. It is assumed
that after the decision of how much to spend out of total income has been made,
that amount is then distributed between domestic consumption and imports.
The estimated import-demand function for food is

$$TBMFOODR = -.131047 - .630081DEVX1AGR + .0373CER$$
$$(-.6) \quad\quad (-2.0) \quad\quad\quad\quad\quad (2.8)$$

$$- .001652PMFOOD/PX1AG$$
$$(-1.4)$$

$$+ .436891TBMFOODR(-1)$$
$$(1.8)$$

$\bar{R}^2 = .795$
$D.W. = 1.74$

where   DEVX1AGR $= X1AGR - X1AGRLT$
       X1AGRLT   = long-run trend of agricultural value added
       CER   = total real consumption expenditures
       PMFOOD/PX1AG = ratio of import price index to agricultural
                             price deflator

The expenditure elasticity is 1.16, while the price elasticity is -.35. The demand for food imports is shown to be quite inelastic with respect to relative prices. This is consistent with the structuralists' hypothesis that food imports have responded to the gap between the expansion of the urban-sector population and the insufficient increase in domestic agricultural output.[15] The statistical significance of the coefficient for deviation from trend emphasized the importance of the internal food-supply conditions in determining imports.

Imports of Other Consumer Goods. Imports of nonfood consumption goods have been constrained by the capacity-to-import variable. As international reserves approached a low level, the authorities responded by restricting imports. This import category has been more vulnerable to balance-of-payments problems than other import items because import restrictions on raw materials, intermediate inputs, and capital goods could produce serious production bottlenecks and those on food could result in serious political conflicts. The explanatory variables are total real consumption expenditures and the level of real international reserves:

$$TBMCR = .113097 + .005434CER + .047433BOPRSVR(-1)$$
$$(1.8) \qquad (1.5) \qquad (2.9)$$

$$+ .130184DUM72/73 + .237986DUM74$$
$$(4.4) \qquad (6.2)$$

$$\bar{R}^2 = .88$$
$$D.W. = 1.21$$

where   BOPRSVR = level of international reserves in constant 1965 pesos
        CER   = total consumption expenditures

The total expenditures elasticity is .41. The inclusion of the dummy variable DUM72/73 for 1972–1973 reflects the scarcity of consumer goods during the Allende period brought about by production bottlenecks and price controls, which augmented pressures on imports of consumer goods. Immediately after the start of the Pinochet administration, many of the import restrictions, such as tariffs, quotas, and prior deposits, were eliminated; the dummy variable for 1974 DUM74 represents the immediate impact of these new policies on the volume of imports.

Capital-Goods Imports. The equation for capital goods imports for the Gran Mineria and the rest of the economy are linkage equations to the level of gross domestic investment. Most of the capital-goods imports are noncompetitive. Although numerous import-substituting industries have been established in Chile, these have been concentrated in the area of intermediate and finished consumer goods. The linkage of capital-goods imports to gross domestic investment assumes that once producers have determined the level of desired capital stock, the breakdown between domestic and imported components is fixed in the short run by the domestic availability of capital goods. Given the technological requirements for imported capital and the relatively unrestricted access to foreign markets, it could be reasonably assumed that the import of capital goods is not a constraint in determining the level of gross domestic investment. The equations for imports of capital for the Gran Mineria and for the rest of the economy are

$$\ln \text{TBMKGMR} = -.3903 + 1.0553 \ln \text{IBGMR} - .0594 \text{TREND}$$
$$(.98) \quad (10.6) \quad\quad (-2.6)$$

$$+ 1.020 \text{ DUM72/73}$$
$$(2.61)$$

$\bar{R}^2 = .948$
$D.W. = 2.01$

$$\text{TBMKNGMR} = \text{MKCOEF} \times \text{IBNGMR}$$

where   TBMKGMR   = real imports of capital for the Gran Mineria
TBMKNGMR = real non-Gran Mineria imports of capital
MKCOEF   = exogenous coefficient, the fraction of gross domestic investment corresponding to imports of machinery and equipment

Raw-Materials Imports. The largest component of imports corresponds to intermediate goods: raw materials and finished goods, which are combined with capital and labor to produce domestic value added. The principal items selected from the total are fuels and intermediate inputs for the Gran Mineria. As in the case of capital goods, the specification of some of the equations assumes a fixed-coefficient technology derived from domestic availability.

The demand for fuel imports is a derived demand equation. It is thus a function of industrial output and relative prices, and actual imports depend on a partial adjustment process to the desired level. The estimated demand equation is

$$\text{TBMFUELR} = -.06689 + .0216 \text{X2R} - .0006 \text{PMFUEL/PGDP}$$
$$(-1.52) \quad (3.99) \quad\quad (1.4)$$

$$+ .4107 \text{ TBMFUELR}(-1)$$
$$(2.83)$$

$\bar{R}^2 = .637$
$D.W. = 2.46$

The price elasticity of demand is -.06; this reflects the limited potential for substituting foreign oil by domestic output of fuel and, to a lesser extent, the very restricted possibilities for substituting other sources of energy or increasing the efficiency of production technology.

The equation for other intermediate imports is determined through production technology in the same way as fuel imports. The relationship between gross output and inputs is approximated by a linear regression between real other intermediate inputs and real non-Gran Mineria gross domestic product:

$$TBMINOTR = -.476 + .0645GDPNGMR$$
$$(-2.8) \quad (7.6)$$

$$\bar{R}^2 = .793$$
$$D.W. = 1.05$$

Exports. Exports of agricultural goods have been competitive in world markets. The export equation is a supply relation between the relative price and real exports. The price term is defined as the export price relative to the agricultural-sector price deflator, and it measures the return from producing for the export market relative to the domestic market. The coefficients of the lag structure for the explanatory variable are assumed to have declining geometric weights. Assuming an infinite lag and applying the Koyck transformation yields the following estimated equation:

$$TBEAGR = .02742 + .0001PEAG/PXIAG + .5363TBEAGR(-1)$$
$$(1.9) \quad (1.8) \quad (3.0)$$

$$- .02302DUM71/73$$
$$(-3.8)$$

$$\bar{R}^2 = .64$$
$$D.W. = 2.2$$

The short-run price elasticity is .13, for the long-run it is .28. The dummy variable DUM71/73 represents the effects of the Allende government's foreign policy on exports and the need to divert agricultural exports to the domestic market to alleviate some of the food-supply shortages existing during that period.

Other exports of goods include noncopper minerals and manufactured goods. These are a function of world demand, which has been measured by an index of gross domestic product for OECD countries. Specifying a partial adjustment process results in the following equation:

$$TBEOTR = .25268 + .001355PRODEV + .40226TBEOTR(-1)$$
$$(2.0) \quad (1.2) \quad (2.03)$$

$$- .16274DUM72/73$$
$$(-2.5)$$

$$\bar{R}^2 = .424$$
$$D.W. = 1.96$$

Other Components of the Balance of Payments. Imports and the export of services are exogenous in the model. There are no published data on freight and insurance costs that separate how much corresponds to exports and how much to imports of goods. Factor services are made up of remittances of dividends on foreign investment and interest payments on the foreign debt. Payment of interest is determined as a function of the level of medium- and long-term debt lagged one period. Transfers are comprised of such miscellaneous items as personal and government transfers to and from abroad and special drawing rights (SDR) allocations, and these are exogenous in the model.

The exchange rate in Chile is determined by the monetary authorities. Whenever there is an excess demand for foreign exchange, the monetary authorities must sell the foreign exchange to the banking system or impose restrictions on imports.[16] If they cannot meet the demand by drawing on international reserves, they must seek short-term financing, paying higher interest rates. In view of the restrictions on the exchange rate and domestic interest rates, capital flows in the balance of payments are more responsive to monetary policy than interest-rate differentials.

Since the medium- and long-term capital flows are determined primarily through various policy decisions on such items as the exchange rate and government borrowing abroad, these are specified as exogenous variables in the model.

Overall Price Determination

The overall price level is one of the crucial variables in the model. Its importance is due to the widely fluctuating and, at times, immense rates of change as well as the dependence of many other variables on prices. The inflation equation explains the variations in the GDP price deflator. Because of the inadequate data on the explanatory variables for price determination by sector, individual sectoral price deflators are linked directly to the GDP deflator. This method has some validity in that many prices, such as nominal wages, prices of services, and prices of foreign exchange, are in some way indexed to the general price level.

The traditional approach to the standard price equation is based on the theory of the firm.[17] With perfect competition, profits are maximized by setting the price of output equal to the marginal cost. If short-run marginal costs consist entirely of labor, the price is set equal to marginal labor costs.

Under imperfect competition, marginal revenues equal output prices modified by a markup factor. The preceding conditions can be summarized as

$$MR = \left(1 + \frac{1}{e}\right)P$$

$$MC = \frac{W}{Q/L} = \frac{W}{MPP_L}$$

where P is the price of output, e is the demand elasticity, W is the wage per worker, and MPP_L the marginal physical product of labor.

In order to combine the two expressions and solve explicitly for the price level, it is necessary to adopt an approximation to the term MPP_L. Since the overall production technology for the noncopper sectors is characterized by a CES production function, substituting the expression for MPP_L from the CES function and simplifying the two expressions yields the following relation between price and unit labor costs:

$$P = \sigma ULC$$

where σ represents the fractional rate of mark up.

This expression binds producers to a neoclassical world, where unit labor costs are equal to the increase in total variable costs from the last person hired and to the value of labor productivity embodied in the last unit produced. The real-world implications are that producers set prices by applying a markup factor to unit labor costs, which represent an approximation to marginal labor costs.

The markup factor is determined by production-function parameters and the demand elasticity. In some models, this coefficient is allowed to vary with the rate of capacity utilization. Producers are more likely to raise their markup whenever they experience a strong upsurge in demand for their products, so variations in unit costs and capacity utilization are the principal determinants of variations in prices as specified in this model. Most models are based on the same price-determination mechanism with varying degrees of sophistication.

In developing countries, especially those with highly unstable rates of inflation, the approach to price determination takes a broader perspective. Because of the distortions arising from a large government sector with persistently high deficits, constraints to output, and undeveloped financial markets, the micro theory of market demand and supply is not sufficient. Too expansive credit policies create excess demand and, hence, more inflation. Excess expansion of demand is a principal cause of persistent inflation. Ad hoc price controls distort relative prices, which leads to a misallocation of resources and more inflation. Developing countries' economies do not have the structural flexibility to respond to changes in the composition of demand. Any significant changes in demand lead to bottlenecks in the supply of output and inputs. With the development of the industrial sector, the growth in demand for food from the urban population has met with insufficient supply form the agricultural sector, and as a result, prices increase.[18] Or, if food prices are regulated by government decree, food supply will be further curtailed.

In deriving the specification of the price equation, elements from these broader approaches were introduced. This permitted a partial evaluation of the relative merits of these theories. The price deflator is made a function of the following variables:

$$\ln \text{PGDP} = \underset{(1.98)}{.16385 \ln \text{WULC123}} + \underset{(15.89)}{.66548 \ln \text{PBMG}} + \underset{(3.12)}{.14911 \ln \text{FM1/Y}}$$

$$+ \underset{(9.77)}{.46068 \ln \text{GDPNGMCAPUT}}$$

$\bar{R}^2 = 1.00$

$D.W. = 2.35$

where WULC123   = unit labor costs for the economy
        PBMG   = import price index (incorporates exchange-
                               rate changes)
        FM1/Y   = money supply (M1) divided by real output
        GDPNGMCAPUT = rate of capacity utilization

The estimation shows the importance of import prices in explaining inflation. This variable is defined as the U.S. dollar import prices multiplied by the exchange rate. Although the latter is in turn determined by the difference between domestic and foreign inflation rates, as through the purchasing-power parity formulation, during the period of estimation, the exchange rate responded with much rigidity and considerable lags to the inflation differentials. The rate of capacity utilization is also a significant factor. Structural variables, such as import prices and the rate of capacity utilization, are statistically more significant than the others in explaining fluctuations in the price level.

Government Sector

This section describes the process leading to the surplus or deficit on current account of the public sector. As mentioned earlier, the public sector includes the central government and the decentralized agencies, such as public firms and local governments.

The predominant item of current revenues is indirect taxes. Between 1960 and 1976, they averaged about 47 percent of total current revenues. This is a characteristic feature of the tax system in many developing countries. Because of the costs involved in implementing direct tax-collection schemes and the extent of tax evasion, indirect taxes have been favored as a reliable instrument of fiscal policy. In the model, indirect taxes are determined as an identity of an implicit tax rate times the current value added at factor costs. The tax rate is assumed to be an exogenous policy variable.

Direct taxes are divided into direct corporate and personal income taxes. Corporate taxes are a function of a tax base that is defined as nonwage income, including Gran Mineria nonwage income (since corporate taxes include proceedings from the Gran Mineria). Personal income taxes depend on personal income net of transfers. A 2-year average of this variable is used in order to account for the lags in payments, and a time trend is introduced as a measure of the increase in efficiency in revenue collection. The estimated equations are

$$\ln \text{TXDCORPN} = \underset{(-35.4)}{-2.34934} + \underset{(8.4)}{.8098} \text{LOFPN}$$

$$+ \underset{(1.6)}{.1876} \text{LOFPN}(-1) - \underset{(-5.7)}{.6167} \text{DUM71/73}$$

$\bar{R}^2 = .970$
$D.W. = 1.81$

$$\ln \text{TXINCOMN} = \underset{(33.2)}{-4.147} + \underset{(20.9)}{.94554} \ln \text{YPNTRNAV} + \underset{(3.90)}{.11231} \text{TREND}$$

$\bar{R}^2 = .997$
$D.W. = 1.57$

The indirect and direct tax equations establish a linkage between the copper micro sector and the government. In the case of indirect taxes, the effects are measured indirectly through total nominal GDP at factor prices, whereas in the latter case, they are incorporated explicitly through nonwage income for the Gran Mineria. During the period of the Allende administration, many private firms were nationalized. These comprised the large banks, manufacturing establishments, and other strategic sectors. The change in status from private to public firms was reflected in a reduction in direct tax revenues. After the initiation of the Pinochet government in the end of 1973, many of these firms were returned to the private sector. A dummy variable has been introduced in the equation for direct taxes to account for these extraordinary political events.

The other endogenous components of current revenues are employers' and employees' social security contributions. The Social Security Administration has operated with some success in spite of the inflationary environment. A large portion of its funds goes into mortgages to finance construction. The contributions are a function of a 2-year average of wages and salaries. The equation for employees includes a time trend for the gradual increase in participation by the workforce. The equations are

$$\ln \text{ESCTOTN} = \underset{(-40.36)}{-1.6997} + \underset{(116.5)}{1.0585} \ln \text{WGSTOTNAV}$$

$\bar{R}^2 = .999$
$D.W. = 1.89$

$$\ln \text{WESCN} = \underset{(67.22)}{-2.5986} + \underset{(76.3)}{.98036} \ln \text{WGSTOTNAV} + \underset{(3.99)}{.03205} \text{TREND}$$

$\bar{R}^2 = 1.00$
$D.W. = 1.44$

Between 1960 and 1976, government purchases of goods and services and transfers to individuals have made up on average 84 percent of total current expenditures. The other items are subsidies and transfers abroad. Real purchases of goods and services are explained through a behavioral equation; the other items are assumed exogenous, determined through the political process. The main variables affecting purchases of goods and services are current revenues in constant pesos and population. In preparing the budget, policymakers usually face a dilemma between the need for fiscal stimulus and the inflationary impact of an unbalanced budget; thus anticipated revenues serve as a constraint on expenditures. The population variable, however, reflects the need for expansion of infrastructure as well as social services. The lag between the proposal, approval, and execution of the budget is introduced in the specification through a partial adjustment mechanism:

$$CEGR = \underset{(-1.47)}{-1.1434} + \underset{(2.69)}{.10167GVRVR1} + \underset{(1.27)}{.18631NP1} + \underset{(2.84)}{.55490CEGR(-1)}$$

$$+ \underset{(3.17)}{.19586DUM71/72}$$

$\bar{R}^2 = .981$

$D.W. = 2.1$

The elasticities with respect to real current revenues are .23 in the short run and .51 in the long run. These values are not surprising in view of the rigidity of the budgetary process. Approved expenditures are usually renewed automatically, and political commitments to some sectors prevent any significant changes in their respective allocations.

The government current-account surplus/deficit is determined as the balance between government revenues and expenditures on the current account. In the national income accounts, there is no information on public investment; thus it is not possible to calculate the overall government surplus or deficit. This is an important element in the determination of the monetary base. The government surplus/deficit on current account has been employed as the linkage from this sector to the expansion of the monetary base and to gross domestic investment.

Employment and Other Factors of Production

This section presents a brief discussion of the employment, the real wage, and the nonwage income equations. The demand for labor is derived from the production process (described in a previous section on supply constraints). Since labor is an essential input into production, the employment equation is an integral part of the mechanism that determines overall supply constraints. Actual employment is a function of a partial adjustment to the demand for labor:

$$\text{NPENGM} - \text{NPENGM}_{-1} = \alpha(\text{NPENGM}* - \text{NPENGM}_{-1})$$

The demand for labor NPENGM* is calculated as the ratio of desired output to an output-per-worker condition determined by the optimization behavior of firms.[19] Thus,

$$\text{NPENGM}* = (\text{GDPNGMR1}* \times 1000)/(\text{GDPNGM}/\text{NPEEST})$$

where GDPNGM/NPEEST is the inverse of the demand for labor per unit output as derived from the first-order conditions of profit maximization. This condition is similar to the one for the aggregate production function:

$$\text{GDPNGM}/\text{NPEEST} = \text{EXP}[.73771 + .174848 \ln \text{W}/\text{PGDPNGM}$$
$$+ .514968 \ln \text{GDPNGM}/\text{NPE}(-1)]$$

The estimated equation is

$$\text{NPENGM} = 28.166 + .54765\text{NPENGM}* + .45101\text{NPENGM}(-1)$$
$$\quad\quad\quad\;\; (.19)\quad\;\; (2.8)\quad\quad\quad\quad\quad\; (2.5)$$

$\bar{R}^2 = .961$
$D.W. = 1.73$

The traditional approach to the determination of nominal wages is based on the Phillips curve analysis, where wages react to disequilibrium in the labor market. The greater is the excess demand for labor, the greater is the increase in wages. The variable used to measure the demand-supply disequilibrium is the rate of unemployment. Changes in the price level are also included in the Phillips-curve approach because of the importance of cost-of-living clauses and money illusion in wage negotiation.

For some developing countries with an excess supply of labor, the Phillips-curve analysis is not as relevant as for industrial countries. In Chile, wage determination occurs through a somewhat different process. With the high rates of inflation, nominal wages have been indexed to some definition of the rate of inflation for a previous period. A reasonable approach to wage determination would be to explain fluctuations in the real wage. By assuming profit maximization, the optimization rule states that the real wage be equal to the productivity of labor. This is a long-run consideration that occurs after producers have adjusted to their desired levels of output and factor inputs as determined through profit maximization. In addition, real wages depend on current rates of inflation in the short run if the readjustments in the money wages fall short of the actual inflation. The real-wage equation for non-Gran Mineria is

$$W123NGMR = -1.0800 + 55450Q/LNGM - .00126PGDP*$$
$$(-1.5)\quad\quad (5.7)\quad\quad\quad\quad (-4.67)$$

$$+ .57860DUM71/72$$
$$(4.40)$$

$\bar{R}^2 = .845$
$D.W. = 2.66$

The dummy variable DUM71/72 represents the income-redistribution schemes of the Allende administration in favor of wage earners. The current inflation rate PGDP* is a proxy variable for the extent of unanticipated price changes. The coefficient for inflation is significantly different from 0, supporting the notion that in higher inflationary environment, the wage readjustments tend to fall short of the actual rates of inflation.

The Monetary Sector

Since interest rates in Chile have been fixed by the monetary authorities, there is no possibility for the market-clearing mechanism to operate through adjustments in the interest rates. The high degree of uncertainty about the value of money makes the opportunity cost of holding cash balances very high. The linkage between the money and the goods markets is established through the effects of money supply on prices. This section describes the determination of the money supply.

The basic money supply identity is

$$M1 = multiplier \cdot MBASE$$

Any increase in monetary assets of the banking system will result in an increase of the money supply by a factor greater than 1. The multiplier effect is caused by the expansion of demand deposits as banks lend a proportion of the original deposits. The size of the multiplier depends on a number of institutional variables, the most important of which is the reserve-requirement ratio. The particular relation for the banking system in Chile is

$$FMPLY = 1.9750 - .0109RRCB + .5535DUM73$$
$$(28.36)\ (-7.78)\quad\quad (3.41)$$

$\bar{R}^2 = .813$
$D.W. = 1.72$

where FMPLY = money multiplier
      RRCB = reserve-requirement ratio
      DUM73 = unusually high reserve ratio during 1973, perhaps due to
               changes in accounting

The greater is the proportion of demand deposits that banks must hold as reserves, the less impact an expansion in the monetary base will have on the money supply.

The reserve-requirement ratio is established by the monetary authorities. It is an instrument of monetary policy used to stabilize the overall price level. Raising the reserve ratio offsets the inflationary impact from an expansion of credit by the banking system. During periods of large government deficits, the monetary authorities attempt to control price increases by raising the reserve-requirement ratio. This behavior is reflected by the following estimated equations:

$$RRCB = 8.8382 + 5.3639 \ln PGDP + 44.031 DUM71/73$$
$$(1.23) \quad (5.19) \qquad\qquad (5.48)$$

$$\bar{R}^2 = .80$$
$$D.W. = 1.73$$

By definition, the monetary base is equal to the sum of Central Bank credit to the public and the private sectors, the level of international reserves, and other net monetary assets and liabilities. The credit to the public sector is estimated by the following equation:

$$DFCRCBGR = 3.9233 - 1.9672 GVSNINTR + 6.3434 DUM73$$
$$(4.31) \quad (-2.25) \qquad\qquad (2.12)$$

$$\bar{R}^2 = .583$$
$$D.W. = .834$$

where DFCRCBGR = real flow of credit to the public sector
      GVSVNINTR = real savings by the public sector on current accounts.
      DUM73   = dummy for Allende government's unusually large
                    borrowing from the central bank

This equation establishes an important link between the government deficit and the price level, and

$$FMBASE = BOPRSVN + FCRCBGN + FMBASEOTHER$$

where FMBASE   = monetary base, end of period stock in current pesos

BOPRSVN   = level of international reserves, end of period stock in current pesos

FCRCBGN   = Central Bank credit to the government, end of period stock in current pesos

FMBASEOTHER = other net assets of the Central Bank, end of period stock in current pesos

The preceding discussion has highlighted the theoretical foundations of the model. These behavioral equations characterize the underlying economic decision-making process. Some of the economic identities, which are necessary to complete the linkages within the model, are presented in the Appendix A. The test of the model's behavioral consistencies will be the topic of the following chapter. Afterwards, some applications will be made to evaluate the sensitivity to different assumptions about the copper-sector variables and to test the hypothesis of export-price instability.

Notes

1. The more recent works are Jere R. Behrman (1977) and Vittorio Corbo Liori (1974). Behrman's model disaggregates production activity by sectors; in addition, various supply constraints are introduced. Corbo Liori's model is not as detailed, but it does link successfully supply and monetary considerations in price determination. The specification of the present model has been influenced by the research of Behrman and Corbo Liori, although the objectives are different. Behrman makes useful application of the model to evaluate government policies.

2. As indicative of the importance of the public-sector investment, Bitar (1974) presents some data for 1960–1970 where the private sector had planned 700 investment projects with a value of US$146 million and the public sector 27 projects with a value of US$363 million.

3. For a useful discussion of linkages between copper and the other sectors, see Ffrench-Davis and Tironi (1977).

4. In each case, the real value of inputs is equal to production times a coefficient. The latter is an exogenous variable defined as production in metric tons divided by the value of the corresponding input definition.

5. See Corbo Liori (1974; Appendix) for a discussion of the properties of the CES production function.

6. Income of self-employed would be included as part of nonwage income.

7. See Evans (1969) for a survey of consumption theories.

8. A useful discussion of the micro foundations of the macro consumption function is found in Wallis (1973; chapter 1).

9. The description of the neoclassical theory and the investment lag structure is based on Jorgenson (1963) and Jorgenson and Siebert (1968).

10. In an extensive study of investment behavior in Chile, Behrman (1972) finds support for the putty-putty hypothesis of capital stock.

11. A survey of the theory of inventory investment is presented in a more recent work with a different interpretation of the adjustment process; see Feldstein and Auerbach (1976).

12. Valdes (1973) cites numerous causes for the demise of the land-reform program.

13. The input-output approach to the specification of value-added equations is developed in Behrman and Klein (1970).

14. For instance, in the equation for imports of nonfood consumption goods, one dummy variable is introduced for the Allende government and another for the trade-liberalization policies of Pinochet. The effect of bottlenecks on imports of nonfood consumption goods is represented by the stock of international reserves.

15. See Wachter (1976) for a detailed analysis of the structuralists' hypotheses.

16. This mechanism is closely linked to the government credit policies and the excess supply of money, which is adjusted through the balance of payments.

17. For a thorough discussion of the micro foundations of the aggregate price equation, see Evans (1969) and Eckstein and Fromm (1968).

18. See Wachter (1976) for a discussion of structuralist and monetarist hypotheses.

19. The derivation presented here is similar to that used in specifying the demand-for-labor equation for the Gran Mineria.

4

Simulations with the Econometric Model: Fiscal and Monetary Policies

The purpose of this chapter is to discuss the overall performance of the Chilean econometric model during part of the sample period (1965–1976) as well as its behavior under alternative fiscal and monetary policies. From an interpretation of the results, many insights are revealed about the importance of various structural relationships. The sample-period simulations are very instrumental in evaluating the sensitivity of the model to the dramatic changes that occurred before, during, and after the Allende government. Another interrelated issue is the validity of the various linkages between the copper sector and the rest of the economy that were specified in the individual equations. The importance of these linkages will be investigated in the following chapter, which will deal exclusively with the interactions between copper and the rest of the economy. Although a particular equation may have statistically significant coefficients, the complex interactions within a simultaneous-equation model may yield results that were unanticipated from the partial equilibrium relation implied by the single equation.

A simulation of the complete model consists of solving for the values of all endogenous variables for the given values of the exogenous and lagged endogenous variables. This is represented by a system of n equations in n unknowns. However, because of nonlinearities of some of the equations, it is not possible to use matrix inversion. The solutions to the system are calculated through the Gauss-Seidel method. Given an initial value, the system iterates until successive values converge within a specified criteria. In addition, the solutions are dynamic, in that lagged endogenous variables appearing in an equation for each period are set equal to the values obtained from the system solution for previous periods. This is an essential element in analyzing the intertemporal as well as secondary effects of an initial "shock" to the system.

In the 12-year base simulation, the official commercial-bank exchange rate and the real flow of credit from the Central Bank to the government are assumed exogenous even though each one is explained by an equation. For the alternative policy simulations, these variables will be endogenous.[1] The exchange rate has been highly responsive in the medium and long run to variations in the ratio of domestic to foreign prices. Any fluctuation in exchange rates will affect all sectors of the economy. Credit from the Central Bank to the government, which is assumed exogenous in the base simulation and endogenous in the policy simulations, plays an important role in the implementation of fiscal and monetary policies.

71

Control Simulation, 1965–1976

The results of the base simulation appear quite reasonable. Table 4–1 shows the actual value, the dynamic simulation value (hereafter denoted as the control solution), and the percentage deviation between the two. Overall accuracy of the model can be measured by the percentage error of gross domestic product, as derived from the dynamic simulation, with respect to the historical values. For the period 1965–1976, the mean absolute percentage error is 2.6 and the root mean squared percentage error is 3.1. This is a particularly good score in light of the radical changes in the 1971–1976 period.

The error statistics for the principal variables are presented in table 4–2. Between 1971 and 1973, the model underestimates actual GDP, although the percentage errors are small. This pattern is explained in part by the quality of the data for that period. Because of the extensive nationalization, many production units were transferred to the public sector, and in that somewhat confused state of affairs, a series of information gaps was created. With the very high rates of inflation, imposition of price controls on many items on the consumer price index, and widespread use of the black market, published price data may not reflect the actual price variations. In spite of these data limitations, the model was able to explain the turning points of the economy for those years. During the politically stable pre-Allende period (1965–1970), the mean absolute percentage error is only 1.27, which compares quite favorably with previous econometric models of Chile of the pre-1971 vintage.

For private-consumption expenditures, the mean absolute percentage error (MAPE) is 3.73. During the Allende period, when significant changes occurred in the income distribution, the percentage errors for private consumption are somewhat above the average for the 1965–1976 period; this is due in part to the real-wage equation, which does not capture the full extent of the increase in income as represented by the dummy variable. Real government consumption expenditures have a MAPE of 4.61. Other aggregate demand variables include non-Gran Mineria gross fixed investment, 5.71; imports of goods and services, 3.47; and exports of goods and services, 2.54. For noncopper investment, the MAPE for the pre-Allende period is 4.25. The difference between the percentage error for the pre-Allende period and for the whole period might be an indication that this variable is more sensitive to the political instability as well as an indication of the negative effects on business-investment decisions of uncertainty and lack of confidence associated with the sudden drastic changes during the period 1971–1976.

In the production account, the MAPEs for primary-, secondary-, and tertiary-sector value added are 3.67, 3.63, and 1.97, respectively. Within the secondary sector, the MAPE for value added by the Gran Mineria is 8.4, which is a very satisfactory performance for that micro sector. For the agricultural sector, the percentage deviations between 1971 and 1973 are 0.93, 10.6, and 0.04,

Table 4-1
Control Simulation, 1965–1976

Date	Actual	Estimated	Residual	Percent Residual
Variable graphed: BOPBN$ (Balance-of-Payments Surplus/Deficit)				
1965	46.70	141.16	−94.46	100.00
1966	119.60	195.53	−75.93	−63.48
1967	−23.40	−99.74	76.34	100.00
1968	117.90	267.34	−149.44	100.00
1969	174.50	254.72	−80.22	−45.97
1970	113.50	145.38	−31.88	−28.09
1971	−299.80	−307.23	7.43	−2.48
1972	−229.00	−233.57	4.57	−2.00
1973	−111.90	−141.30	29.40	−26.28
1974	−45.10	−231.59	186.49	100.00
1975	−274.60	−93.64	−180.96	65.90
1976	455.00	634.95	−179.95	−39.55
Variable graphed: CEGR (Real Government Consumption)				
1965	1.99	1.96	0.03	1.49
1966	2.19	2.12	0.07	3.27
1967	2.19	2.28	−0.09	−4.28
1968	2.30	2.40	−0.10	−4.42
1969	2.43	2.56	−0.13	−5.19
1970	2.59	2.78	−0.19	−7.37
1971	2.83	3.08	−0.25	−8.87
1972	2.97	3.25	−0.28	−9.28
1973	3.07	3.20	−0.13	−4.23
1974	3.30	3.18	0.12	3.69
1975	3.07	3.05	0.02	0.57
1976	3.25	3.17	0.09	2.63
Variable graphed: CEPR (Real Private Consumption)				
1965	13.36	13.76	−0.40	−3.01
1966	14.78	14.36	0.42	2.83
1967	15.30	14.64	0.66	4.30
1968	15.79	15.09	0.70	4.41
1969	16.22	15.93	0.29	1.79
1970	16.87	16.32	0.54	3.22
1971	18.56	17.51	1.05	5.67
1972	19.67	18.46	1.21	6.15
1973	18.78	18.15	0.63	3.38
1974	18.55	18.21	0.33	1.80
1975	16.75	15.91	0.84	5.00
1976	16.30	15.66	0.64	3.95
Variable graphed: CUPRODGM (Production of Copper, Gran Mineria)				
1965	479.20	543.77	−64.57	−13.47
1966	536.80	557.28	−20.48	−3.82
1967	536.40	560.19	−23.79	−4.43
1968	519.30	587.71	−68.41	−13.17
1969	540.20	602.64	−62.44	−11.56
1970	540.70	611.75	−71.05	−13.19

Table 4-1 continued

Date	Actual	Estimated	Residual	Percent Residual
Variable graphed: CUPRODGM (Production of Copper, Gran Mineria)				
1971	571.30	589.27	−17.97	−3.15
1972	592.60	604.41	−11.81	−1.99
1973	615.30	608.42	6.88	1.12
1974	762.90	647.21	115.69	15.16
1975	682.30	659.23	23.07	3.38
1976	854.10	864.22	−10.12	−1.18
Variable graphed: CUPRODPMM (Production of Copper, Medium- and Small-Scale Mines)				
1965	106.10	105.95	0.15	0.14
1966	99.90	117.84	−17.94	−17.96
1967	123.80	123.93	−0.13	−0.11
1968	137.70	129.67	8.03	5.83
1969	147.90	136.18	11.72	7.92
1970	150.90	140.07	10.83	7.18
1971	137.00	135.66	1.34	0.98
1972	124.20	131.14	−6.94	−5.59
1973	120.10	135.11	−15.01	−12.49
1974	139.20	142.81	−3.61	−2.59
1975	146.00	139.22	6.78	4.64
1976	151.10	138.92	12.18	8.06
Variable graphed: FM1 (Money Stock, M1)				
1965	2.25	2.54	−0.29	−13.04
1966	3.06	4.28	−1.22	−39.76
1967	3.74	4.37	−0.64	−17.02
1968	5.42	7.50	−2.08	−38.31
1969	7.74	11.17	−3.43	−44.25
1970	12.09	16.19	−4.10	−33.88
1971	25.84	25.70	0.14	0.54
1972	70.48	56.88	13.61	19.30
1973	361.79	526.97	−165.19	−45.66
1974	1255.71	1097.51	158.20	12.60
1975	4735.59	9127.36	−4391.77	−92.74
1976	15362.63	22492.12	−7129.49	−46.41
Variable graphed: GDPNGMCAPUT (Rate of Capacity, Utilization)				
1965	87.54	85.92	1.62	1.85
1966	90.97	88.20	2.77	3.05
1967	90.16	89.16	1.00	1.11
1968	92.14	89.35	2.79	3.03
1969	94.00	93.50	0.50	0.53
1970	94.37	94.52	−0.16	−0.17
1971	100.52	98.85	1.67	1.66
1972	100.60	98.84	1.76	1.75
1973	92.95	92.64	0.31	0.34
1974	93.41	92.15	1.27	1.36
1975	84.70	82.02	2.68	3.16
1976	85.85	83.19	2.66	3.10

Table 4-1 continued

Date	Actual	Estimated	Residual	Percent Residual
Variable graphed: GDPR (Real Gross Domestic Product)				
1965	18.75	18.60	0.15	0.82
1966	20.07	19.69	0.38	1.89
1967	20.55	20.20	0.36	1.73
1968	21.17	20.63	0.54	2.57
1969	21.90	22.04	-0.14	-0.62
1970	22.69	22.69	0.00	0.00
1971	24.44	23.46	0.98	3.99
1972	24.42	23.60	0.81	3.34
1973	23.53	23.10	0.43	1.84
1974	24.87	23.53	1.34	5.37
1975	22.06	20.94	1.12	5.06
1976	22.96	21.98	0.98	4.29
Variable graphed: GVEXPN (Current Government Expenditures)				
1965	4.11	4.02	0.09	2.27
1966	6.04	5.97	0.07	1.18
1967	7.95	8.00	-0.05	-0.62
1968	10.73	11.18	-0.45	-4.24
1969	14.92	15.83	-0.91	-6.06
1970	23.71	23.69	0.02	0.07
1971	39.90	37.92	1.98	4.96
1972	75.58	67.73	7.85	10.38
1973	307.39	329.39	-22.00	-7.16
1974	2354.61	2241.35	113.27	4.81
1975	9046.74	10219.64	-1172.89	-12.96
1976	30722.19	29535.62	1186.57	3.86
Variable graphed: GVRVN (General Government, Current Revenues)				
1965	4.95	4.84	0.10	2.12
1966	7.65	7.40	0.25	3.32
1967	10.00	9.98	0.01	0.12
1968	13.37	13.89	-0.52	-3.85
1969	20.20	21.59	-1.38	-6.84
1970	31.41	33.94	-2.53	-8.05
1971	39.34	39.30	0.04	0.10
1972	64.19	61.03	3.16	4.93
1973	254.14	265.81	-11.67	-4.59
1974	2920.40	2400.34	520.06	17.81
1975	9955.90	11280.03	-1324.13	-13.30
1976	36888.05	37001.09	-113.04	-0.31
Variable graphed: GVSVN (General Government, Current Savings)				
1965	0.84	0.83	0.01	1.35
1966	1.61	1.43	0.18	11.30
1967	2.04	1.98	0.06	2.98
1968	2.65	2.71	-0.06	-2.29
1969	5.28	5.76	-0.48	-9.04
1970	7.70	10.25	-2.55	-33.07

Table 4-1 continued

Date	Actual	Estimated	Residual	Percent Residual
Variable graphed: GVSVN (General Government, Current Savings)				
1971	−0.56	1.38	−1.94	100.00
1972	−11.38	−6.70	−4.68	41.15
1973	−53.25	−63.58	10.33	−19.39
1974	565.79	159.00	406.79	71.90
1975	909.16	1060.39	−151.24	−16.64
1976	6165.86	7465.47	−1299.61	−21.08
Variable graphed: IBNGMR (Real Gross Fixed Investment Excluding Gran Mineria)				
1965	2.83	2.70	0.13	4.59
1966	2.86	2.74	0.12	4.03
1967	2.75	2.72	0.03	1.00
1968	2.73	2.70	0.03	1.24
1969	2.88	2.80	0.08	2.84
1970	3.16	2.79	0.37	11.80
1971	3.29	2.98	0.31	9.47
1972	2.91	2.87	0.04	1.45
1973	2.82	2.70	0.12	4.28
1974	2.75	2.47	0.28	10.19
1975	2.02	1.85	0.17	8.55
1976	2.03	1.85	0.19	9.13
Variable graphed: MGSR (Real Imports of Goods and Services)				
1965	2.37	2.43	−0.06	−2.69
1966	3.00	3.00	0.01	0.24
1967	2.91	3.02	−0.11	−3.68
1968	3.16	3.11	0.05	1.56
1969	3.53	3.40	0.13	3.62
1970	3.68	3.82	−0.14	−3.92
1971	3.68	3.80	−0.12	−3.33
1972	3.80	3.89	−0.09	−2.32
1973	4.04	3.91	0.13	3.22
1974	4.15	3.89	0.26	6.35
1975	3.60	3.23	0.37	10.23
1976	3.07	3.09	−0.01	−0.44
Variable graphed: NPENGM (Number of Persons Employed Excluding Gran Mineria)				
1965	2482.75	2460.93	21.83	0.88
1966	2540.03	2516.27	23.76	0.94
1967	2627.03	2558.45	68.58	2.61
1968	2671.46	2589.92	81.54	3.05
1969	2691.52	2679.51	12.01	0.45
1970	2746.40	2735.37	11.03	0.40
1971	2879.97	2790.91	89.06	3.09
1972	2968.83	2829.37	139.46	4.70
1973	2984.12	2801.38	182.73	6.12
1974	2912.30	2889.96	22.33	0.77
1975	2808.10	2818.52	−10.42	−0.37
1976	2907.64	2823.48	84.17	2.89

Table 4-1 continued

Date	Actual	Estimated	Residual	Percent Residual
Variable graphed: PGDP* (Percent Change of GDP Price Deflator)				
1965	34.95	34.78	0.17	0.50
1966	30.35	37.27	-6.92	-22.80
1967	28.27	24.04	4.22	14.94
1968	30.93	39.15	-8.22	-26.59
1969	40.64	39.65	0.98	2.42
1970	38.79	35.41	3.38	8.72
1971	23.57	25.13	-1.56	-6.60
1972	85.36	63.17	22.19	26.00
1973	426.69	499.83	-73.14	-17.14
1974	653.66	593.40	60.27	9.22
1975	391.14	516.10	-124.96	-31.95
1976	234.69	156.85	77.85	33.17
Variable graphed: W123NGMN (Wage per Worker, Excluding Gran Mineria)				
1965	2.82	2.82	-0.00	-0.12
1966	3.91	4.07	-0.16	-4.03
1967	4.82	5.16	-0.34	-7.14
1968	6.60	7.20	-0.59	-9.01
1969	9.40	10.53	-1.13	-12.02
1970	14.11	14.49	-0.38	-2.71
1971	21.14	22.05	-0.91	-4.29
1972	40.40	35.08	5.32	13.17
1973	142.74	176.99	-34.26	-24.00
1974	1104.13	943.58	160.54	14.54
1975	4898.70	5194.41	-295.70	-6.04
1976	17043.11	16612.69	430.42	2.53
Variable graphed: X1AGR (Real Value Added, Primary Sector)				
1965	1.78	1.87	-0.09	-4.98
1966	1.93	1.90	0.03	1.61
1967	2.06	1.94	0.12	5.86
1968	2.11	2.00	0.11	5.36
1969	1.92	2.05	-0.13	-6.62
1970	2.04	2.10	-0.06	-2.99
1971	2.17	2.15	0.02	0.93
1972	2.08	1.86	0.22	10.60
1973	1.78	1.78	-0.00	-0.04
1974	2.08	2.03	0.04	2.06
1975	2.16	2.13	0.02	1.15
1976	2.21	2.17	0.04	1.83
Variable graphed: X2CONTR (Real Value Added, Construction)				
1965	1.00	0.98	0.02	1.96
1966	0.96	0.94	0.02	1.80
1967	0.92	0.96	-0.04	-4.26
1968	0.93	0.95	-0.02	-2.18
1969	1.01	1.05	-0.04	-4.24
1970	1.04	1.01	0.03	2.42

Table 4-1 continued

Date	Actual	Estimated	Residual	Percent Residual
Variable graphed: X2CONTR (Real Value Added, Construction)				
1971	1.15	1.02	0.13	11.55
1972	1.05	1.01	0.03	3.29
1973	0.92	0.88	0.05	5.04
1974	1.11	0.89	0.22	19.69
1975	0.76	0.74	0.03	3.40
1976	0.62	0.59	0.03	5.25
Variable graphed: X2GMR (Real Value Added, Gran Mineria)				
1965	0.88	1.05	-0.17	-19.66
1966	1.04	1.08	-0.04	-4.13
1967	1.08	1.09	-0.02	-1.47
1968	0.98	1.14	-0.15	-15.35
1969	1.03	1.14	-0.10	-10.01
1970	0.99	1.13	-0.13	-13.55
1971	0.96	0.97	-0.01	-0.65
1972	0.95	0.98	-0.03	-3.24
1973	1.06	1.02	0.04	3.74
1974	1.31	1.07	0.24	18.45
1975	0.90	0.85	0.05	5.39
1976	1.30	1.37	-0.07	-5.42
Variable graphed: X2MFGR (Real Value Added, Manufacturing)				
1965	4.57	4.35	0.22	4.82
1966	4.96	4.66	0.30	5.95
1967	5.10	4.75	0.35	6.85
1968	5.22	4.78	0.45	8.60
1969	5.38	5.25	0.13	2.48
1970	5.45	5.40	0.05	0.94
1971	6.20	5.87	0.33	5.25
1972	6.37	5.98	0.39	6.11
1973	5.96	5.73	0.23	3.79
1974	5.90	5.42	0.48	8.20
1975	4.29	4.07	0.22	5.15
1976	4.58	4.27	0.31	6.72
Variable graphed: X2MINNGMR (Real Value Added, Mining Excluding Gran Mineria)				
1965	0.89	0.83	0.06	6.73
1966	0.89	0.90	-0.01	-0.84
1967	0.87	0.96	-0.10	-11.13
1968	1.00	1.03	-0.03	-2.97
1969	1.20	1.09	0.11	9.38
1970	1.26	1.15	0.11	8.91
1971	1.34	1.21	0.13	9.64
1972	1.30	1.28	0.03	1.96
1973	1.23	1.34	-0.11	-9.33
1974	1.34	1.39	-0.05	-3.90
1975	1.62	1.43	0.19	11.79
1976	1.59	1.44	0.15	9.29

Table 4–1 continued

Date	Actual	Estimated	Residual	Percent Residual
Variable graphed: X2R (Real Value Added, Secondary Sector)				
1965	7.63	7.50	0.13	1.66
1966	8.15	7.90	0.25	3.12
1967	8.32	8.11	0.22	2.63
1968	8.46	8.24	0.22	2.60
1969	8.95	8.91	0.05	0.52
1970	9.09	9.10	−0.01	−0.13
1971	10.04	9.50	0.54	5.38
1972	10.12	9.70	0.42	4.14
1973	9.63	9.42	0.20	2.12
1974	10.18	9.24	0.94	9.23
1975	8.09	7.55	0.55	6.77
1976	8.63	8.18	0.45	5.23
Variable graphed: X2UTR (Real Value Added, Public Utilities)				
1965	0.29	0.29	−0.00	−0.42
1966	0.31	0.31	−0.01	−2.49
1967	0.36	0.34	0.02	6.05
1968	0.33	0.36	−0.03	−8.59
1969	0.33	0.39	−0.05	−16.30
1970	0.35	0.41	−0.07	−19.25
1971	0.39	0.43	−0.04	−10.52
1972	0.45	0.45	−0.00	−0.30
1973	0.46	0.46	0.01	1.19
1974	0.52	0.47	0.05	9.27
1975	0.52	0.46	0.06	11.75
1976	0.54	0.51	0.03	6.05
Variable graphed: X3R (Real Value Added, Tertiary Sector)				
1965	9.34	9.22	0.12	1.23
1966	9.98	9.89	0.09	0.94
1967	10.17	10.15	0.02	0.15
1968	10.59	10.38	0.21	1.99
1969	11.03	11.09	−0.06	−0.54
1970	11.57	11.50	0.07	0.59
1971	12.22	11.81	0.41	3.35
1972	12.22	12.04	0.18	1.43
1973	12.13	11.90	0.23	1.90
1974	12.61	12.26	0.35	2.80
1975	11.81	11.27	0.54	4.59
1976	12.12	11.63	0.49	4.07

respectively; the unusually high value for 1972 is explained, in part, by the drastic changes and political confrontations that characterized the land-reform program.

The solution values of gross domestic product for each simulation are determined by the sum of primary-, secondary-, and tertiary-sector value added. This

Table 4-2
Chile, Mean Average Percent Error (MAPE) for Principal Variables during
Historical Simulation, 1965-1976
(*percent*)

	MAPE
Agriculture value added	3.67
Secondary sector value added	3.63
Tertiary sector value added	1.97
Private consumption	3.73
Public consumption	4.61
Non-Gran Mineria investment	5.71
Imports of goods and services	3.47
Exports of goods and services	2.54
Public sector, current savings	25.34
Money: M1 (N)	33.09
GDP price deflator (N)	6.41
Gran Mineria, value added	8.46

Note: All variables are expressed in constant 1965 pesos except those marked by (N), which
are in nominal terms.

should be equal to total aggregate demand, which represents the other half of
the balance sheet for the economy. Since total value added in the model is determined by the production side of the accounts, there is a statistical discrepancy
between the sum of aggregate demand components and sectoral output. In the
base simulation, the average deviation between the two was 0.39 and 0.59 percent in constant and current pesos, respectively. For each of the government-
policy and copper-sector simulations, the statistical discrepancy between the two
definitions of GDP show the same pattern as in the control solution. Whenever
there is some minor discrepancy between the control and the alternative simulation, the adjustment is assumed to occur through the demand variables (stock of
inventories).[2]

The monetary- and price-sector variables did less satisfactory than the real-
sector variables. This is due to some extent to the sensitivity of the monetary-
base components to any fluctuation in the underlying variables. The level of
international reserves varies with the surplus or deficit in the balance of payments, which, in turn, is determined as the difference between flows in the
current and capital accounts. The impact of any errors would be intensified by
taking their difference, and this volatility would be reflected in the monetary-
base identity. The MAPE for the implicit GDP price deflator is 6.41, and for the
monetary base, 30.12. When the size of the average percentage errors for the
monetary variables is compared with the average rate of inflation between 1965
and 1976 (168.2 percent), the model appears to fit the actual values reasonably
well. A MAPE of 30 percent or even 50 percent for a nominal variable is still

indicative of the success of the model in tracing the acute fluctuations in the overall price level. The stability of the estimated equation is also reflected in the fact that none of the variables shows a consistent build-up of simulation error.

Policy Analysis with the Econometric Model, 1965-1976

This section describes a series of applications of the model to analyze the consequences of alternative government policies within the context of the macro as well as the micro copper sector. The simulations discussed in this chapter test the behavioral characteristics of the model: whether the individual variables are behaving properly, and whether the values for the multipliers are reasonable. In addition to establishing numerous linkages between policy variables and the rest of the economy, especially between the copper sector and the macro model, the simulations are very useful in showing the importance of these linkages. In the dynamic context, this is a useful device for studying cyclical behavior and the underlying growth process. The three basic areas for these simulations are fiscal policy, monetary policy, and factor-incomes policy.

Within the sphere of fiscal policy, two basic scenarios are considered: an increase in real government expenditures, and an increase in public investment. These are very relevant instruments in achieving growth, redistribution of income, and price stability. The monetary-policy simulation consists of a variation in the reserve-requirement ratio. Factor-income policy is represented by changes in the real wage that could result from legislation governing price indexing or labor negotiations with the public sector. Finally, the copper-sector simulations presented in the following chapter analyze the myriad of interrelations with the macro model with special attention to the indirect effects on noncopper income and employment. Changes in copper production, prices, exchange rate, and gross investment are relevant issues to be considered. The length of a simulation varies according to the amount of time required for the simulation path to return to the path of the control solution. Most of the experiments depart from a 1-year increase or decrease of an exogenous variable with respect to its control-solution value; a sustained increase or decrease is used in some cases to analyze the accumulation effect and whether it is damped or intensified by the interactions within the system.

An Increase in Government Expenditures

A 10 percent increase for 1 year in real government consumption provides the basic scenario of expansionary fiscal policy. The increase is calculated with respect to the values for the control simulation. Since the objective of this simulation experiment is to study the relative magnitude of the effects of an increase in government expenditures on the rest of the model, the initial increase could also have been 15 or even 25 percent and still be considered as a reasonable scenario.

Real government consumption is determined through a behavioral equation, so th. 't the increase is made by an additive adjustment to the equation. For the initial s'mulation run, both the exchange rate and the real flow of credit from the Central Bank to the government are made endogenous. Production of copper by the Gran Mineria and the medium- and small-scale mines, as well as the nominal wage for the Gran Mineria, is made exogenous (they are set equal to the figures from the control simulation). This latter assumption does not alter the basic conclusions of the simulations, which focus on the stabilization or the long-run growth path of the noncopper sectors. Even if there is some price effect on the domestic cost component of copper production, the output response to this particular increase would be minimal given the low price elasticity. However, copper production is an important element in the balance of payments, and it would be difficult to identify the linkages of fiscal policy with respect to balance of payments and other components of the model if there are concomitant changes in copper production.

Table 4-3 shows how the 1-year increase in government consumption has affected the principal variables in the first and subsequent years until the system returns to its original path, as characterized by the control simulation. A measure of the overall response of the economy to an initial shock is obtained through the size of the multiplier effect on GDP. There are two types of multiplier effects discussed in the simulation studies for the Chilean economy: the impact multiplier, and the intermediate-run path of the economy back to the original "equilibrium" values. The former measures the change in GDP in the first year of the simulation divided by the change in government consumption. The latter is illustrated through a comparison of the differences between the values of the GDP for the alternative simulation and the control simulation as a result of the initial change in government consumption.

The size of the impact multiplier is calculated by the following expression:

$$\frac{GDPR_{65}^{d} - GDPR_{65}^{c}}{CEGR_{65}^{d} - CEGR_{65}^{c}} = \frac{\Delta GDPR\ 65}{\Delta CEGR_{65}} = \frac{.2}{.196} = 1.02$$

where $GDPR_{65}^{d}$ = GDPR for 1965, "disturbed" path
       $GDPR_{65}^{c}$ = GDPR for 1965, control path
       $CEGR_{65}^{d}$ = CEGR for 1965, "disturbed" path
       $CEGR_{65}^{c}$ = CEGR for 1965, control path

The impact multiplier is a relevant index when comparing alternative policy simulations. It is a measure of the relative magnitude of the deviation of the

Table 4-3
Fiscal Policy Simulation, a 1-Year Increase in Government Consumption
(percent deviation from control solution)

	1965	1966	1967	1968	1969	1970
Production (constant 1965 pesos)						
Gross domestic product	1.10	.36	.05	.10	.07	-.01
Agriculture	0.0	-.36	.12	.05	.14	.10
Secondary sector	1.18	.25	.03	.10	.07	-.01
Construction	.40	-.27	.08	.01	.03	-.04
Manufacturing	1.70	.59	.08	.15	.10	-.04
Gran Mineria	.59	-.41	-.32	-.37	.06	.02
Non-Gran Mineria mining	0.0	.06	.04	.06	.08	.09
Public utilities	.39	.30	.19	.15	.12	.07
Tertiary sector	1.25	.59	.05	.10	.05	-.02
Demand (constant 1965 pesos)						
Private consumption	1.43	.12	-.40	-.38	-.26	-.08
Public consumption	9.78	5.33	2.64	1.49	.78	.38
Non-Gran Mineria investment	.47	-.27	.25	.10	.05	.21
Imports of goods and services	1.17	.68	.21	.10	-.05	-.02
Rate of capacity utilization	.91	.05	-.08	-.05	-.15	-.03
Income and employment (constant 1965 pesos)						
Per capita net of tax wages and salaries	1.08	-.06	-.31	.07	.16	-.02
Per capita net of tax nonwage income	2.11	-.07	-.60	-.75	.60	-.02
Employment, non-Gran Mineria (000 persons)	.45	0.0	-.08	-.11	-.10	0.0
Public sector						
Current revenues (current pesos)	8.62	5.18	3.31	1.04	.24	.07
Current expenditures (current pesos)	9.89	4.43	3.00	.97	.48	.25
Current savings (in 1965 pesos)	-6.83	4.53	1.03	.73	-.30	-.43
Flow of credit to public sector (in 1965 pesos)	10.04	-10.64	-2.11	-.46	-1.98	.08
Money and prices						
Balance-of-payments surplus or deficit (mill. US$)	-6.40	-3.59	-2.05	-.41	-.26	+.14
Money supply (M1)	8.94	3.70	3.39	1.98	1.40	1.07
GDP price deflator (1965 = 100)	10.05	3.61	3.52	.59	-.10	.09

"shocked" solution from the control path. Some policy variables have a greater immediate impact or maintain the increase in activity for a longer period of time. The latter effect is demonstrated through a comparison of the results for subsequent periods.

In the first year, the increase in government consumption has a direct impact on value added for the secondary and tertiary sectors, (X2R, X3R). Within the secondary sector, the increase occurs mainly in manufacturing. The increase in the level of activity leads to more imports and a deterioration in the balance of payments by an amount of US$9.0 million. The level of international reserves diminishes, implying a reduction of the monetary base and the possibility of future restraints on imports. The increase in government expenditures results in an increase in the current account deficit (or decrease in savings), requiring additional financing by the Central Bank through an expansion of domestic credit to the public sector. Since the money supply is determined endogenously in the system, the credit expansion increases the money supply in spite of the contractionary effect exerted by the depletion of international reserves. The increase in the money supply combined with an increase in the rate of capacity utilization results in higher prices. The GDP price deflator is 10 percent higher in the first year of the simulation than in the control solution.

The inflation generated by the expansionary policy checks further increases of expenditures and, hence, of total output expansion in the economy. In the foreign sector, imports of goods and services increase at a slightly higher rate than the GDP during the first three years. This may be a response to an increase in aggregate demand in excess of the potential domestic supply. Since the official exchange rate has been adjusted to maintain a constant purchasing-power parity of the currency, very little of the increase in imports could be ascribed to changes in the relative prices. Thus the deterioration in the balance of payments occurs as a result of a growth in imports of intermediate as well as finished goods; the latter are necessary to narrow the gap between domestic demand and supply.

One of the reasons for the increase in the overall price level is that in the supply-determined sectors, productive potential is almost constant. Primary-sector value added shows no response in the first year, as the structuralists would expect. Because of the small relative price elasticity (.24) and the lag between the change in prices and the increase in agricultural output, any changes in demand would have little effect on supply in the short run. There is almost no impact on gross domestic investment, which is essential for growth of the economy. With a growth of only 0.47 percent in investment in the first year and a decrease in three out of the other four years, there is insufficient accumulation of capital stock. The capital is important in creating sufficient capacity to meet the expansion in demand, as well as to replace the worn out with more efficient equipment to improve productivity and raise real wages. The increase in aggregate demand with productive capacity relatively constant in the short run leads to a depletion of inventories without any subsequent changes in the production sectors.

During the first year, real government savings on current account decrease by 6.8 percent, but increase during the next 3 years. Real flow of credit from the Central Bank to the government is higher by 10 percent the first year and declines for the following years. The increase in purchases of goods and services by the government is offset by greater revenues. Indirect taxes rise by 10.7 percent in nominal terms, or a negligible increase in real terms, while the other tax components in nominal terms fail to rise faster than the price level. While real government revenues fail to match the increase in expenditures during the first year, they subsequently show greater buoyancy than real current expenditures, leading to a small improvement in real current-account savings.

The expansionary fiscal policy has some effects on the allocation of resources and the distribution of income. The first is characterized by a change in the aggregate demand composition. The average share of government consumption out of total output in the control solution is 11.35 percent, and for private consumption it is 72.71 percent. For the "disturbed" simulation path, the relative shares of government and private consumption are 11.67 and 72.65 percent, respectively. The lack of response by the production sector results in some reallocation from private- to public-sector consumption through an induced inflation process. The changes in factoral distribution of income are not as pronounced as in the case of resource allocation; however, there is some redistribution of income in favor on nonwage income. Real per capita wage and nonwage incomes (net of taxes) increase by 1.0 and 2.0 percent, respectively, during the first year. This is the outcome of two price-related phenomena: first, wages are not adjusted sufficiently for the increase in prices; and second, the higher prices are usually associated with greater profits for the entrepreneur.

In summarizing the principal conclusions of the simulation, the main point is that an expansionary fiscal policy such as an increase in government purchases of goods and services would have a low multiplier effect because of the capacity constraints in the economy. The initial financing of the expenditures leads to an expansion of the money supply and to upward pressures on prices. Increased demand without any additions to productive capacity also leads to higher prices; output from existing capacity is insufficient to satisfy demand, resulting in a depletion of inventories. Finally, the increase in demand for goods by the government leads, in the end, to a reallocation from private to public consumption.

In order to magnify some of the results of the preceding simulation, real government consumption was increased by 10 percent with respect to the control simulation each year from 1965 to 1976. The economy goes through a similar adjustment process and returns to the control simulation path after 6 years. Table 4-4 presents some of the important variables. The GDP price deflator is significantly higher than in the previous simulation. For the first 2 years, the price deflator is 10 percent above the control solution in the case of

Table 4-4
Fiscal Policy Simulation, A Sustained Increase in Government-Consumption Expenditures
(percent deviation from control solution)

	1965	1966	1967	1968	1969	1970
Production (constant 1965 pesos)						
Gross domestic product	1.10	.39	1.53	.83	.32	.44
Agriculture	0.0	-.38	-.18	.13	.30	-.81
Secondary sector	1.18	.26	1.70	.69	.28	.36
Construction	.40	-.28	1.19	-.36	-.30	-1.09
Manufacturing	1.70	.67	2.49	1.18	.42	.59
Gran Mineria	.59	-.32	.38	-.44	.36	.67
Non-Gran Mineria mining	0.0	.06	.02	.16	.16	.13
Public utilities	.39	.31	.75	.69	.46	.25
Tertiary sector	1.25	.64	1.72	1.05	.35	.73
Demand (constant 1965 pesos)						
Private consumption	1.43	-.12	1.23	-.12	-.33	.66
Public consumption	9.78	9.74	10.39	10.65	8.70	9.23
Non-Gran Mineria investment	.47	-.49	1.37	-.41	-.16	-1.14
Imports of goods and services	1.17	.65	1.93	.93	.01	.45
Rate of capacity utilization	.91	-.16	1.24	-.20	-.53	-.39
Income and employment (constant 1965 pesos)						
Per capita net of tax wages and salaries	1.08	-.04	1.49	.85	-.79	-1.24
Per capita net of tax nonwage income	2.11	-.06	2.72	.36	1.38	2.38
Employment, non-Gran Mineria (000 persons)	.45	-.14	.52	-.11	-.31	-.34
Public Sector						
Current revenues (current pesos)	8.62	10.73	8.70	2.72	22.68	20.36
Current expenditures (current pesos)	9.89	10.52	8.58	5.68	23.05	16.92
Current savings (in 1965 pesos)	-6.83	.59	2.31	-10.13	-7.11	6.25
Flow of credit to public sector (in 1965 pesos)	10.04	.15	-10.38	44.96	38.08	-48.31
Money and prices						
Balance-of-payments surplus or deficit (mill. US$)	-6.40	3.44	-19.31	-3.64	.03	-4.32
Money supply (MI)	8.94	12.73	5.35	15.56	38.24	20.28
GDP price deflator (1965 = 100)	10.05	10.95	6.73	.43	30.97	20.78

the sustained increase in expenditures as compared with 10 and 3.6 percent in the case of a 1-year increase. Gross domestic investment is as equally unresponsive to the sustained increase in government consumption as with the 1-year increase. The reallocation of resources in favor of public-sector consumption is more significant, because its share in GDP increased from 11 to 13 percent while the share for private consumption declined from 74 to 72 percent.

A second variant of the government-consumption simulation assumes that the 1-year 10 percent increase in expenditures is accompanied by an increase in domestic taxes and external financing during the first year. Domestic financing is in the form of an increase in indirect taxes equal to the increase in government consumption. The external financing is through the long-term capital account in the balance of payments; specifically it is obtained through an increase in disbursements of medium- and long-term loans to offset the initial deterioration of the balance of payments.

An examination of the results in Table 4–5 will reveal little if any substantial changes. The impact multiplier is only .56, with minimal increases in GDP in 2 of the following 3 years. The reduction in the impact multiplier is due to the method of financing the increase in expenditures. The use of indirect taxes neutralizes some of the stimulative effect on production and capacity utilization that would have occurred without the tax burden.

This negative effect operates through the income accounts. National income is obtained from GDP after deducting indirect taxes, while nonwage income is calculated as a residual after subtracting wage income from nonwage income. Thus in the first year of the simulation with domestic financing, nonwage income declines from 8.1 to 8.0 million current pesos and the real per capita nonwage income net of taxes declines about 2 percent, in contrast to an increase of 2 percent in the previous case. This is reflected in a smaller increase in aggregate demand, as demonstrated by the rate of capacity utilization, which decreases very slightly with the domestic and external financing in contrast to an increase of about 1 percent in the case of no exogenous financing.

By financing the expansion in government consumption with an increase in indirect taxes, the government does not have to rely on credit from the monetary authorities. The increase in the real flow of credit from the Central Bank to the government is now only 8 percent compared with 10 percent without taxes. This reduced expansion of the money supply results in less inflation, with the GDP price deflator dropping during the first year by about 1 percent with tax financing, whereas in the previous simulation the deflator was 10 percent higher.[3]

An Increase in Government Investment

The evaluation of government-consumption expenditures policy has centered on the inflationary impact and the reallocation of resources. The following simulation traces the effects of an expansion in public investment, and the discussion

Table 4-5
Fiscal Policy Simulation, A 1-Year Increase in Government-Consumption Expenditures Accompanied by Foreign and Domestic Financing
(percent deviation from control solution)

	1965	1966	1967	1968	1969	1970
Production (constant 1965 pesos)						
Gross domestic product	.58	.30	-.80	.09	-.14	-.08
Agriculture	0.0	.03	-.10	-.66	.07	-.13
Secondary sector	.37	.14	-1.06	-.01	-.33	-.13
Construction	-1.09	-.76	-1.68	-.84	-.86	-.25
Manufacturing	.66	.41	-1.39	.10	-.25	-.04
Gran Mineria	.27	-.11	.25	.31	.23	0.0
Non-Gran Mineria mining	0.0	-.10	-.19	-.42	-.53	-.62
Public utilities	.13	.16	-.33	-.35	-.21	-.14
Tertiary sector	.87	.48	-.72	.31	-.02	-.02
Demand (constant 1965 pesos)						
Private consumption	.86	.25	-.78	.18	-.02	0.0
Public consumption	11.28	5.78	1.90	1.61	.89	.44
Non-Gran Mineria investment	-.74	-.51	-1.87	-.51	-.70	.05
Imports of goods and services	.81	.41	-.92	.25	.13	.14
Rate of capacity utilization	-.14	.03	-1.31	-.23	.13	.18
Income and employment (constant 1965 pesos)						
Per capita net of tax wages and salaries	1.15	.63	-1.47	-.60	.19	-.01
Per capita net of tax nonwage income	-2.14	.19	-.16	1.82	.88	-.23
Employment, non-Gran Mineria (000 persons)	-.24	-.23	-.85	-.40	-.18	-.13
Public Sector						
Current revenues (current pesos)	2.42	.04	9.05	2.98	.13	.05
Current expenditures (current pesos)	4.31	2.79	7.61	1.15	.54	.28
Current savings (in 1965 pesos)	-5.39	-11.32	1.52	10.51	-.83	-.67
Flow of credit to public sector (in 1965 pesos)	8.01	33.75	-6.47	-40.12	2.00	.74
Money and prices						
Balance-of-payments surplus or deficit (mill. US$)	3.04	-2.53	8.73	-1.22	-.88	-1.78
Money supply (MI)	1.41	12.62	14.74	-2.31	-2.05	-1.34
GDP price deflator (1965 = 100)	-1.37	-.04	13.25	.14	-.16	.20

will focus on some of the significant differences between these two types of expansionary policies: consumption versus investment. Government-investment expenditures increase in the first year by an amount equal to 10 percent of the value of gross fixed investment for the control simulation. Since most development programs have involved large expenditures for the development of infrastructure, the initial outlays are followed by additional induced expenditures as the investment projects are carried to completion. The dynamics of the investment function that links current investment expenditures to a previous investment decision have been described in the section dealing with the description of the model. In terms of the specifics of the model, this simulation shows the impact of a government-investment project with an initial price tag of .27 million pesos in 1965.

This simulation assumes that the investment project has a gestation period of 1 year. However, the increase in productive capacity and, hence, realized output does not occur until subsequent years. The initial investment expenditures could have been distributed over 2 or even 3 years, but this would not have altered significantly the impact multiplier nor the time path of the economy in relation to the path of the control solution.

Using the same notation as in the previous section, the simulation yields the following impact multiplier:

$$\frac{\Delta GDPR65}{\Delta IBNGMR65} = \frac{.97}{.44} = 2.20$$

These results show that the overall consequences for the economy are significantly more favorable with government investment than with consumption expenditures. Enlarged productive capacity makes possible larger output increases for a substantially longer period of time. The "disturbed" solution returns to the original path of the control solution after 9 years, versus 5 years in the previous simulation. Table 4-6 presents the results for some of the important variables, which are analyzed in the following discussion.

The initial response in production occurs in the secondary and tertiary sectors, which increase by 7.7 and 4.2 percent, respectively. If this is expressed in terms of an impact multiplier, the figures would then be 2.1 for the secondary sector and 1.4 for the tertiary sector, which indicates significantly greater response in production to an increase in investment through the creation of productive capacity than through consumption expenditures. Within the secondary sector, construction value added rises immediately by 14.8 percent as part of investment activity; and manufacturing, by 10.1 percent as a response to greater overall demand in the economy. Because of a low price elasticity and the lagged response to investment expenditures in infrastructure, agricultural value added begins to show significant response after 2 to 3 years. The upsurge in production is reflected in greater employment, with an increase of 2.9 percent in the first

Table 4-6
Fiscal Policy Simulation, A 1-Year Increase in Government Investment
(percent deviation from control solution)

	1965	1966	1967	1968	1969	1970	1971	1972	1973
Production (constant 1965 pesos)									
Gross domestic product	5.19	5.08	7.10	6.17	7.17	4.91	3.43	2.01	1.64
Agriculture	0.0	1.23	1.88	2.80	3.33	4.38	4.39	5.14	5.38
Secondary sector	7.72	7.16	9.89	8.63	9.59	6.58	4.99	2.85	2.30
Construction	14.84	12.48	14.69	12.54	10.83	6.52	4.89	.76	-.68
Manufacturing	10.06	9.10	12.18	10.54	11.22	7.19	4.55	1.74	.78
Gran Mineria	-.75	-.38	.74	.07	.62	-.21	-.49	-.27	.33
Non-Gran Mineria mining	0.0	2.13	4.41	6.90	8.85	10.43	11.24	11.47	11.07
Public utilities	2.94	4.56	6.21	7.01	7.57	7.16	6.46	5.43	4.57
Tertiary sector	4.17	4.16	5.87	4.87	5.92	3.69	2.00	.85	.56
Demand (constant 1965 pesos)									
Private consumption	3.69	4.23	7.17	6.10	8.30	5.18	3.56	2.26	2.08
Public consumption	1.95	2.82	3.97	3.77	4.94	3.99	2.14	1.58	.82
Non-Gran Mineria investment	16.37	13.73	17.10	13.56	13.02	7.96	5.53	.67	-.66
Imports of goods and services	7.44	6.05	7.20	5.19	4.96	2.20	.28	-1.58	-2.00
Rate of capacity utilization	6.14	5.20	6.09	4.68	5.30	2.19	.89	-.95	-.87
Income and employment (constant 1965 pesos)									
Per capita net of tax wages and salaries	7.56	6.94	8.62	8.19	8.47	6.18	3.92	2.15	1.84
Per capita net of tax nonwage income	4.80	4.33	9.03	5.87	7.50	3.81	-.70	1.18	.77
Employment, non-Gran Mineria (000 persons)	2.93	3.42	4.78	4.94	6.03	5.08	4.49	3.46	3.22
Public Sector									
Current revenues (current pesos)	-4.36	-8.19	-10.14	-9.70	-13.81	-11.32	-4.32	-3.58	-8.82
Current expenditures (current pesos)	-4.47	-5.73	-6.92	-6.23	-10.05	-6.63	-1.39	-1.81	-3.88
Current savings (in 1965 pesos)	8.00	-4.92	-6.36	-10.43	-1.90	-7.15	-123.53	-19.67	-26.98
Flow of credit to public sector (in 1965 pesos)	-15.72	17.14	22.26	44.11	18.50	55.70	9.96	4.72	4.56
Money and prices									
Balance-of-payments surplus or deficit (mill. US$)	-40.96	-31.29	-72.64	-20.74	-23.73	-22.47	-2.29	9.35	23.53
Money supply (M1)	-22.6	-24.06	-32.26	-23.74	-33.36	-19.96	-3.75	-.84	-6.71
GDP price deflator (1965 = 100)	-10.94	-14.15	-18.10	-15.48	-22.53	-16.21	-4.75	-5.49	-8.25

year, or an average of 3.4 percent for the period 1965-1973. These figures demonstrate how an efficient public-investment program generates employment: first, as labor requirements for the investment projects; and second, through the multiplier effect of additional aggregate demand that is accompanied by an increase in productive capacity.

Public investment brings about a redistribution of income slightly in favor of the nonwage earners. For the control solution, the average proportion of the wage bill out of total factor income excluding the Gran Mineria was 52.65 percent for the period of duration of the initial shock, while the one for the "shocked" solution is 52.40 percent. This change is not very significant; however, it does point to some potential bias of public investment. The causality from investment to income redistribution would differ from the preceding case if the government applied higher taxes on nonwage income and used the proceeds to increase its transfers to individuals as well as nominal wages to public employees. This scheme may not pose a threat of demand-pull inflation if the increase in investment would lead to an expansion of production capacity.

During the period when GDP increases with respect to the control solution, the GDP price deflator is decreasing; and in the last 3 years, it begins to rise significantly. Although the drop might be explained in part by diminishing supply constraints, the movements in the price level reflect some of the limitations of the model. Since the real flow of credit from the Central Bank to the public sector has been specified as a function of the public savings on current account, the model does not calculate the possible Central Bank financing required for public investment and hence does not register any impact in terms of additions to the money supply. The other limitation is perhaps more characteristic of econometric models of developing countries. This is the nonexistence of an endogenous capital account in the balance of payments. The increase in gross domestic investment—through its import-intensive component of machinery and equipment—leads to an increase in imports of capital goods. As a result, there is a deterioration of the balance of payments of almost $58 million in the first year, with similar deficits occurring during the following 6 years. These declines in international reserves would not be realized if long-term capital inflows responded to the needs of the public sector for additional financing.

A more realistic assessment of the impacts of public investment is obtained by an alternative scenario in which the real flow of credit to the public sector is increased by the amount of the investment expenditures. The results for this alternative are presented in table 4-7. With explicit financing of the investment expenditures, the impact multiplier is somewhat lower: 2.07 versus 3.59 without financing, whereas the difference is not as significant for the dynamic multiplier effect of the two simulations. The differences can be explained by the GDP price deflator, which increases by 6.2 percent in the first year when the investment is financed through the real credit expansion but declines by 10.9 percent in the first simulation without the explicit financing. Even though in both cases, after

Table 4-7
Fiscal Policy Simulation, A 1-Year Increase in Government Investment Accompanied by Domestic Financing
(*percent deviation from control solution*)

	1965	1966	1967	1968	1969	1970	1971	1972	1973
Production (constant 1965 pesos)									
Gross domestic product	3.03	5.36	6.29	6.06	6.54	5.43	3.50	2.20	1.89
Agriculture	0.0	.33	1.98	2.52	3.06	4.11	4.41	5.08	5.25
Secondary sector	5.08	7.52	8.71	8.56	8.86	7.33	5.14	3.28	2.72
Tertiary sector	1.97	4.59	5.19	4.76	5.30	4.16	2.01	.89	.73
Demand (constant 1965 pesos)									
Private consumption	1.39	4.89	5.50	6.20	7.40	4.87	3.77	2.18	1.31
Public consumption	.23	2.77	3.33	3.24	4.30	3.95	2.24	1.62	1.00
Non-Gran Mineria investment	12.72	13.96	15.89	14.26	12.78	9.69	5.34	1.88	.34
Imports of goods and services	4.54	6.78	6.56	5.14	4.70	2.79	.52	-1.16	-1.62
Rate of capacity utilization	2.76	5.36	5.57	4.62	5.00	2.92	.71	-.66	-1.16
Income and employment (constant 1965 pesos)									
Per capita net of tax wages and salaries	4.25	6.62	7.67	7.03	7.73	6.57	4.17	2.76	2.89
Per capita net of tax nonwage income	3.27	6.36	5.19	4.41	5.61	3.03	-.20	.29	1.55
Employment, non-Gran Mineria (000 persons)	1.26	3.09	3.97	4.32	5.16	4.75	4.00	3.14	2.71
Public Sector									
Current revenues (current pesos)	7.36	-2.32	-6.07	-4.66	-9.32	-9.63	-3.96	-2.27	-5.88
Current expenditures (current pesos)	3.06	-4.15	-4.68	-3.60	-7.46	-6.40	-1.34	-1.25	-2.72
Current savings (in 1965 pesos)	20.80	18.25	1.14	.78	3.47	-1.96	-75.80	-12.19	-17.92
Flow of credit to public sector (in 1965 pesos)	-5.23	-16.59	37.02	38.64	16.59	44.66	12.48	4.94	3.68
Money and prices									
Balance-of-payments surplus or deficit (mill. US$)	-24.95	-34.69	-67.26	-20.61	-22.51	-27.69	-3.32	6.65	18.92
Money supply (MI)	-1.37	-22.04	-23.61	-19.59	-27.45	-18.59	-1.04	2.01	-3.17
GDP price deflator (1965 = 100)	6.02	-11.01	-12.67	-9.77	-17.27	-15.46	-4.61	-3.97	-6.42

the first year of the simulation, overall prices are lower than in the control solution, the decline is significantly less for the simulation with domestic financing.
In this last case, the lower GDP growth is reflected in lower import requirements
and, hence, less of a strain on the balance of payments and the level of international reserves. The relatively higher price level in the first year of the simulation with financing checks the increase in aggregate demand; as a consequence,
value added in the tertiary sector, which is highly sensitive to demand, expands
by half the rate, as in the alternative case. In terms of income distribution, there
are no significant differences between the two simulations.

The effects of investment expenditures are very different from those of
consumption expenditures, as seen by their effects on prices, productive
capacity, and employment. The results of the simulations indicate some need for
concomitant balance-of-payments policy to prevent the deterioration of international reserves. The drawing down of reserves is a consequence of a greater
demand for imports of machinery and intermediate goods brought about by
increased domestic activity.

The size of the investment-induced deficits in the balance of payments
would require borrowing abroad, and if the government is able and willing to
obtain the necessary financing through long-term credits, the resulting foreign
borrowing might further reduce the multiplier effect through import restrictions,
the burden of a larger foreign debt, and increased pressures on the price level.

Monetary Policy

The two instruments of monetary policy that have been specified in the model
are the expansion of the money supply through financing of the government
deficit and the reserve-requirement ratio applicable to commercial banks. The
effects of the former have been analyzed in the discussions of fiscal policy, so
this section will evaluate the impacts of a change in the reserve-requirement
ratio. The small size of financial markets has severely constrained the use of
open-market operations, that is, the purchase or sale of securities in the open
market to control the money supply. The policy simulation will consist of a 10
percent in the reserve ratio with respect to the control-solution values for every
year of the simulation period.

The impact of an increase in the reserve-requirement ratio on the money
supply and the price level in the model are the ones expected from the standard
macroeconomic theory. Table 4-8 presents the results of this simulation for
some of the more important variables. In the first year, there is a decrease in the
money multiplier and a contraction in the money supply. The money multiplier
is calculated through a behavioral equation and is a function of the reserve ratio.
As the monetary multiplier decreases, the money supply contracts by 2.1 percent. The monetary base, however, remains almost constant, since the decline in

Table 4-8
Monetary Policy Simulation, A Sustained Increase in the Reserve-Requirements Ratio
(percent deviation from control solution)

	1965	1966	1967	1968	1969	1970	1971	1972	1973	1974	1975	1976
Production (constant 1965 pesos)												
Gross domestic product	-.07	.01	-.06	-.03	-.07	.42	.13	-.05	.16	-1.02	-.63	-2.13
Agriculture	0.0	.02	-.04	-.05	-.10	-.17	-.15	-.07	-.28	-.22	-.49	-.41
Secondary sector	-.06	-.07	-.22	-.21	-.22	.37	0.0	-.31	-.09	-1.80	-1.51	-3.39
Tertiary sector	-.10	.08	.07	.12	.06	.58	.29	.15	.42	-.57	-.06	-1.57
Demand (constant 1965 pesos)												
Private consumption	.67	.72	.30	.16	.67	.71	.57	.57	.95	.42	.48	-1.94
Public consumption	.15	.12	.12	.09	.12	.26	.50	.27	0.0	-.14	-.09	-.92
Non-Gran Mineria investment	-.22	-.46	-.61	-.79	-.79	-.96	-.51	-.99	-1.09	-4.44	-5.35	-8.95
Rate of capacity utilization	.27	.34	.38	.31	.56	.24	.74	.45	.79	-.26	-.15	-1.72
Income and employment (constant 1965 pesos)												
Per capita net of tax wages and salaries	-.14	-.23	.25	.10	-.08	-.34	.61	.03	.92	-1.57	2.67	-3.61
Per capita net of tax nonwage income	.99	.24	-.66	-.40	.26	1.54	.74	1.41	2.77	.64	-1.30	-.22
Public Sector												
Current savings (in 1965 pesos)	-.74	-1.15	-1.53	-1.51	-.73	.74	-67.13	31.39	26.01	-65.77	-63.35	-25.32
Flow of credit to public Sector (in 1965 pesos)	1.77	3.35	5.35	6.15	4.99	-3.95	7.06	7.66	4.40	6.80	9.06	15.36
Money and prices												
Balance-of-payments surplus or deficit (mill. US$)	-.42	-.18	-1.01	.02	-.27	-.79	-1.99	-1.29	-4.54	7.72	15.01	9.06
Money supply (M1)	-2.07	-1.46	-1.46	-.56	-.44	-3.36	-8.25	-7.47	-13.20	-6.12	-23.55	-1.97
GDP price deflator (1965 = 100)	-.66	-.33	-.33	-.26	-.11	-1.78	-5.94	-5.29	-7.71	-6.74	-15.90	-4.57

the level of international reserves is offset by the small increase in Central Bank credit to the public sector. The ability of commercial banks to create money is checked by the higher reserve ratio, while in addition, they must adjust their balance sheets in order to attain the new level of required reserves. With the higher reserve ratio, the overall price level is 0.66 percent lower in the first year with respect to the control simulation. Throughout the simulation period, the money supply is lower than in the control solution, with the exception of 1 year. The price level is also lower, as would be predicted from the standard IS-LM curve macroeconomic analysis.

The reserve-requirements policy has some deleterious effects on the domestic economy that would detract from the positive consequences of price stability. This phenomenon is brought about by a decrease in the real current-account savings of the public sector. During the first 5 years of the simulation, the increase in the government current-account deficit is reflected in a lower level of gross domestic investment. Thus GDP fluctuates around the path of the control solution, with the variations depending on the intensity of the decline in investment relative to the other aggregate demand components. The decrease in the stock of capital affects supply determination by the production sectors, especially agriculture, which has lower values throughout the period versus the control simulation. Private-consumption expenditures increase by .67 percent the first year and continue to expand until the last year. In view of the contraction in productive capacity, the increase in aggregate demand is not accompanied by sufficient response from the production sectors, especially the secondary sector, and as a consequence, inventories are diminished. A part of the excess demand is directed toward imports of goods and services. The rise in imports leads to a deterioration of the balance of payments and a reduction in the monetary base. However, the increase in Central Bank credit to the government that is used to finance the public-sector current-account deficits works against the overall reduction in the monetary base, and in the first year the two effects cancel out.

The impact of an increase in the reserve requirements on income distribution is not clear from the results of the simulation. The cyclical response of the GDP brought about by the decline in real public savings is, in part, the cause of the variations of the ratio of wage to the sum of wage and nonwage incomes (excluding the Gran Mineria) around the path of the control solution. The rate of capacity utilization increases by 0.24 percent in the first year in response to the constraints on domestic productive capacity and is maintained at higher rates until the last 3 years. The results of this simulation can be summarized by two propositions: one is that whenever the monetary authorities adjust the reserve-requirement ratio, some gains are made in terms of price stability; the second may be considered a corollary to the first, in that when the government imposes monetary restraints, these should be accompanied by cutbacks in current expenditures instead of reductions in investment outlays in order to maintain productive capacity in line with the price-induced increases in demand.[4]

Wage-Adjustment Policy

As described previously, the noncopper wage equation was specified as a relation among real wages, productivity, and the inflation rate. The latter variable has a negative coefficient because of the unanticipated price increases. The determination of the level of wages is an important mechanism in the model, affecting employment as well as the overall price level. Since price determination is based in part on the markup model, nominal wages affect the price level through unit labor costs. The purpose of this simulation is to investigate the extent of the linkage between an increase in wages and aggregate demand and the overall price level. The model is subjected to a 1-year 20 percent increase in the nominal wage with respect to the control-solution value with the aim of tracing its dynamic effects on prices and the growth path of the economy. In this case, the impact on GDP from the 1-year increase in wages dies out after 6 years, and on inflation, after the first year. The results for this simulation are shown in table 4-9.

In the initial year, the increase in nominal wages is 26.7 percent, which is higher than the original increase of 20 percent, since the immediate inflationary impact of the wage increase is factored in through further wage increases. On the supply side, the wage increase increases unit labor costs by 22.4 percent. Producers react to the increase in unit labor costs by adjusting two of their decision variables: price and employment. How much of the added labor costs can be passed on to the consumer as higher prices will depend on the demand elasticities that enter in the determination of the markup factor. The employment effect of higher factor costs is a reduction in the demand for labor until the real wage is equal to the marginal productivity. Total employment excluding the Gran Mineria decreases by 1.6 percent.

Since the elasticity of substitution between capital and labor is less than 1 (.175 in the short run), the higher labor cost would result in an increase in the share of wage income. Thus the initial reduction in employment through capital-labor substitution is offset by the income effect on aggregate demand through a higher share of wage income, so that output shows a net positive increase. Real per capita wages and salaries net of taxes, which represent the wage-income variable in the consumption equation, rise by 2.19 percent; however, the nonwage-income variable decreases by 16.8 percent as a result of the higher labor costs and the limited capacity of producers to pass on all the increase in the form of higher prices.

In spite of the opposite effects on wage and nonwage income, the more favorable changes in the former coupled with the greater marginal propensity to consume out of wage income results in a net increase in private-consumption expenditures of 1.95 percent.

The peak in GDP occurs in the second year, and then the difference between the two paths is diminished every period until the seventh, when the stimulative effect of the wage increase dies out. In the second year of the "cycle," invest-

Table 4-9
Wage-Adjustment Policy Simulation, A 1-Year Increase in Nominal Wages
(*percent deviation from control solution*)

	1965	1966	1967	1968	1969	1970
Production (constant 1965 pesos)						
Gross domestic product	.92	2.32	.97	.91	.36	1.11
Agriculture	0.0	-.11	1.14	-.01	.27	.33
Secondary sector	1.03	2.85	1.13	1.24	.38	1.29
Tertiary sector	1.02	2.37	.80	.81	.36	1.10
Demand (constant 1965 pesos)						
Private consumption	1.95	2.41	.85	1.21	1.29	1.56
Public consumption	.39	3.13	1.00	.78	.62	.90
Non-Gran Mineria investment	.59	3.92	1.50	1.55	-.80	-.09
Imports of goods and services	.97	3.12	.69	.59	.03	.49
Rate of capacity utilization	.44	3.37	1.07	.47	.07	.69
Income and employment (constant 1965 pesos)						
Per capita net of tax wages and salaries	21.9	2.10	2.80	1.26	.45	.83
Per capita net of tax nonwage income	-16.80	2.20	-.70	1.10	1.80	2.00
Employment, non-Gran Mineria (000 persons)	-1.60	.25	-.26	-.27	-.22	.25
Public Sector						
Current revenues (current pesos)	5.17	-9.10	-4.80	-.80	-.46	-1.94
Current expenditures (current pesos)	1.89	-7.58	-.87	-.50	-.13	-1.53
Current savings (in 1965 pesos)	16.90	3.10	-18.60	-.37	-.21	1.3
Flow of credit to public sector (in 1965 pesos)	-38.70	-5.90	63.90	2.03	2.00	-6.80
Money and prices						
Balance-of-payments surplus or deficit (mill. US$)	-5.29	-16.06	-7.16	-2.41	-.12	-4.67
Money supply (MI)	-15.90	-26.00	-3.90	-3.60	-2.60	-6.00
GDP price deflator (1965 = 100)	3.4	-18.1	-2.7	-1.8	-1.1	-4.1
Unit labor costs	22.44	-15.36	-2.58	-1.49	-.99	-4.01

ment expenditures increase at a faster rate than total consumption, as would be expected from the accelerator hypothesis. However, a decrease in real government savings started in the third year leads to a drop in the growth of investment (the government component), which, coupled with declining aggregate demand, begins to fall below the control-solution values in the fifth year of the simulation.

Within the production sectors, the response varies between agriculture and the rest of the economy. Because of the lagged supply response in agriculture, there is no change in the first year and a drop of .11 percent in the second year before the increase in productive capacity and the positive relative price effect result in a higher level of output. The increases in value added by the secondary and tertiary sectors are almost the same: 1.03 and 1.02 percent, respectively. Thus total value added in the economy grows by .92 percent with respect to the control solution in the first year. Because of insufficient expansion in productive capacity, the rate of capacity utilization increases by .44 percent.

The combination of an increase in unit labor costs and in the rate of capacity utilization for the non-Gran Mineria sectors outweighs the effect of the first-year drop in the money supply to bring about an increase in the overall price level or GDP deflator of 3.4 percent. The drop in the money supply is due to a reduction in the monetary base, which is attributed to a decrease in the level of international reserves and the Central Bank credit to the public sector. The initial increase in domestic production is made possible in part through imports of intermediate goods, and the induced growth in aggregate demand is also manifested through more imports of consumption goods, including food. Imports of goods and services are .97 percent higher than in the control solution during the first year. The balance of payments registers a deficit of US$7.47 million, which is translated into a net decrease in the level of international reserves held by the banking system. Although there is a positive response in real government-consumption expenditures through higher revenues, real current-account savings increase by 16.9 percent. Current revenues grow by more than current expenditures, since the increase in wages means that more income taxes are collected. The improvement in the government budget results in a decrease in the real flow of credit to the public sector of 38.7 percent, and hence, nominal credit outstanding declines by 11.3 percent. After the first year, the price level begins to decrease along with unit labor costs and the money supply. The latter is affected by the continued deterioration in the balance of payments and diminishing credit from the Central Bank to the public sector.

As the price level continues to fall below the original control values, aggregate demand picks up. This occurs just when the income effect of the initial increase in wage income disappears. The most pronounced effect on income distribution is recorded during the first year. Afterwards there is some redistribution in favor of wage income during the third year, while for some of the other years, nonwage income shows a slight gain in its share. The cyclical response of the economy to the increase in wages is evinced by the reallocation of resources toward consumption goods. The share of total consumption out of GDP in the first and sixth years of the control simulation is 84.0 and 84.6 percent, respectively, with the increase in wages, the respective shares are 84.7 and 84.9 percent. The decision to affect changes in the wage rates must be thought out in terms of price stability as well as the net social benefits derived from the intertemporal reallocation of resources. The latter issue is very complicated because of the sensitivity of the analysis to any arbitrarily chosen social rate of discount.

Conclusions

The study of the various simulations in this chapter has provided useful policy recommendations and has validated the behavioral relationships within a simultaneous-equations model. A noteworthy fact is that by redistributing resources from consumption to investment expenditures, the public sector can strengthen the growth of the economy. The investment multiplier is significantly greater than the consumption multiplier. The monetary and wage-adjustments policies are particularly relevant to the attainment of price stability. The trade-off between monetary stringency or limited wage adjustments in controlling inflation is primarily a political one, although the incomes policy may have an additional effect through price expectations.

The numerical results of the simulation are not a mathematically precise valuation of the effects of various policies; instead, they are indicative of the relative impact on the important variables. This type of analysis is most useful in identifying the problem areas as well as the beneficiaries originating from different government actions.

Notes

1. The equation for determining the exchange rate is based on the theory of purchasing-power parity.

2. The statistical discrepancy in the alternative simulations is within an acceptable margin of error with respect to the control-solution values.

3. These results are consistent with the standard macro theory of the balanced-budget multiplier.

4. Investment expenditures have not been a dynamic factor in the growth of the Chilean economy, so that the public sector should guard against sudden cutbacks in capital outlays.

5

Analysis of the Impact of Copper-Sector Variables

The historical analysis, discussion of the econometric methodology, and evaluation of model performance through simulations of government policies that have been detailed in the previous chapters are combined in this chapter for the purpose of analyzing the interactions between the copper and noncopper sectors of the economy. The format consists of simulations that depict the myriad of interrelations with the macro model with special attention to the indirect effects on noncopper income and employment. There are five central variables within the copper sector: production, prices, investment, exchange rate, and price variability. Production, investment, and the exchange rate are determined by the economic decisions of the copper producers or by government policies aimed at maximizing the participation of the economy in the revenues earned by that sector. The level of the international price of copper is determined by world supply and demand through the London Metal Exchange (LME). Thus prices and price variability are an outcome of fluctuations in the world copper market, which the Chilean producers are unable to control. The first four variables comprise the set of simulations that are discussed in this chapter. The extent of the deleterious effects of export instability on domestic activity is a controversial issue in the literature; and by the application of counterfactual simulation analysis, an attempt is made in this study to clarify some of that confusion for the case of Chile. Since the policy response to copper-price variability differs from that of the first four variables, its impact on domestic activity and policy options will be discussed in the next chapter.

The evolution of the Gran Mineria, from a foreign-owned enclave to the majority participation by the public sector and finally to the nationalization, has raised many questions concerning the net benefits to the Chilean economy of the economic decisions made first by the foreign stockholders, then by a public sector with ownership of the majority of the stock, and finally by a public sector with total ownership and control of the mines, as occured throughout the period 1960–1976. By tracing the effects of variations in each of the five variables on the rest of the Chilean economy, these simulations provide useful information for the development of national copper policies.

In assessing the contributions from changes in the price as well as the quantity of copper, the first two simulations are specified as the control-solution assumptions plus an additional condition: for the first simulation, it is assumed that the LME price of copper is 10 percent higher every year than the corresponding historical value, and for the second, it is assumed that the quantity of copper produced is 10 percent higher every year. By comparing the results of independent simulations for a quantity and a price change, it is possible to quantify the sensitivity of other variables in the model to fluctuations in the copper sector. The third simulation analyzes the benefits from investing in the copper industry. This is evaluated in the context of the alternative of noncopper investment by the public sector, which is described in chapter 4. Finally, the fourth simulation examines the impact on the copper industry from maintaining an overvalued exchange rate for the Gran Mineria. In recent studies this policy has been criticized because of its pernicious effects on copper production, and this chapter attempts to quantify some of these effects through a macro model simulation. The focus of this chapter and the next is on a description of the underlying economic relations as specified within the econometric model, so that they can serve as a basis for the formulation of some copper-policy recommendations in chapter 8.

Increase in Copper Production

The previous simulations in chapter 4 have dealt with government policies mainly within the noncopper sectors of the economy. The results were very helpful in evaluating the linkages between the various sectors of the Chilean economy. They also provided some interesting insights into the pros and cons of the different policy instruments often applied by the government as part of a stabilization program. The following simulations deal with a very strategic sector of the economy with a high policy priority, especially because of its importance within the balance of payments, public-sector budget, and a conglomeration of other variables in determining the potential growth path of the economy. Variations in output or export price of copper have substantial effects on the rest of the economy. Since the model has been specified with special attention to these factors, the copper-sector simulations should bring out in detail most of these linkages. The first simulation assumes an increase in the level of copper production by the Gran Mineria that is 10 percent above the control-solution values. The results are presented in table 5-1.

The 10 percent increase in the level of copper production is almost matched by a 10 percent increase in value added. The proportionality between production and value added is established by the assumption of fixed coefficients for domestic and imported inputs throughout the simulations. Based on the values for value added for the Gran Mineria and the total for the economy, it is possible to calculate the impact multiplier:

$$\frac{\Delta GDPR_{65}}{\Delta X2GMR_{65}} = \frac{.28}{.1} = .280$$

Table 5-1
Copper Sector, A Sustained Increase in Production by Gran Mineria
(percent deviation from control solution)

	1965	1966	1967	1968	1969	1970	1971	1972	1973	1974	1975	1976
Gran Mineria												
Value added (1965 pesos)	10.00	10.06	9.86	10.05	10.82	10.23	9.80	10.67	10.69	9.14	8.46	9.25
Employment (000 persons)	2.80	5.00	6.53	7.70	8.70	9.38	10.00	10.56	10.51	10.87	11.14	11.59
Wage bill (current pesos)	3.48	6.85	3.32	4.33	-.71	-8.49	3.20	24.00	6.87	12.88	14.30	19.25
Nonwage income (current pesos)	15.48	11.15	3.41	5.73	-3.40	-14.90	157.14	-8.13	9.80	8.25	9.37	17.40
Production (constant 1965 pesos)												
Gross domestic product	1.51	2.40	3.25	3.80	5.32	6.48	4.40	3.23	2.29	1.08	.89	.77
Agriculture	0.0	-.02	.12	.43	.39	.94	1.43	.99	1.33	1.16	.96	.74
Secondary sector	2.04	2.76	3.82	4.43	6.41	7.75	5.13	3.65	2.24	.71	.26	.31
Tertiary sector	1.38	2.56	3.40	3.95	5.35	6.49	4.35	3.24	2.47	1.34	1.30	1.10
Demand (constant 1965 pesos)												
Private consumption	1.22	2.09	2.98	3.50	5.98	6.59	4.42	3.78	2.95	1.30	.32	-.06
Public consumption	.36	.89	1.73	1.84	3.16	4.70	2.55	2.06	1.56	1.05	.62	.29
Non-Gran Mineria investment	.45	.91	2.63	2.91	5.44	6.46	3.61	1.16	-1.98	-4.61	-6.59	-7.31
Imports of goods and services	1.55	2.11	3.59	3.84	5.51	6.06	4.21	2.48	1.65	.58	-.14	-1.24
Rate of capacity utilization	.95	1.46	2.87	2.83	5.06	6.00	2.88	1.28	.69	-.17	-.34	-.67
Income and employment (constant 1965 pesos)												
Per capita net of tax wages and salaries	1.95	2.28	3.70	4.10	5.94	6.41	5.33	3.76	1.84	.78	.59	.45
Per capita net of tax nonwage income	2.72	2.91	3.89	4.93	8.52	8.54	4.01	4.12	4.08	.78	-.32	.69
Employment, non-Gran Mineria (000 persons)	.41	.65	1.43	1.68	3.17	4.22	3.28	2.59	2.15	1.39	.92	.54

Table 5-1 continued

	1965	1966	1967	1968	1969	1970	1971	1972	1973	1974	1975	1976
Public Sector												
Current revenues (current pesos)	2.10	1.94	-1.53	-.93	-4.81	-10.79	-3.56	-.31	-.30	.80	2.34	6.95
Current expenditures (current pesos)	.57	.03	-2.16	-1.44	-4.90	-8.50	-1.02	-.69	-.42	.48	1.47	4.21
Current savings (in 1965 pesos)	8.52	10.81	7.25	6.19	8.60	4.85	-81.85	-.28	1.60	4.92	8.65	10.32
Flow of credit to public sector (in 1965 pesos)	-19.37	-30.93	-23.15	-23.00	-51.19	-21.95	8.46	-.01	.28	-.50	-1.14	-7.31
Money and prices												
Balance-of-payments surplus or deficit (mill. US$)	27.72	20.96	10.62	10.16	4.92	-4.54	.47	11.12	45.36	39.02	79.37	19.34
Money supply (MI)	.48	-4.27	-11.50	-11.50	-22.21	-29.50	-8.44	-5.62	-2.80	7.25	7.03	12.19
GDP price deflator (1965 = 100)	.86	-.83	-6.04	-6.04	-12.10	-19.99	-4.66	-3.69	-2.48	.32	1.93	6.87

The dynamic multiplier effect is shown by the variations in GDP, which reaches a peak in 6 years at about 6.5 percent. This pronounced effect is attributed to the direct as well as indirect linkage between the copper sector and the rest of the economy, which is brought out over time through the lagged variables in the equations.[1]

Copper Sector

Within the copper sector, the increase in production brings about an immediate increase in employment. In the first year it increases by 2.8 percent, followed by successively higher rates for the other years in the simulation. The average increase in employment in the Gran Mineria is about 2400 persons per year, which corresponds to an average increase in output of 62,000 metric tons per year. The continuous increases in demand for labor are explained through the nature of the demand-for-labor equation. As the real wage decreases, copper producers increase their demand for labor, and by specifying a VES production technology, the lower capital intensity results in a more labor-intensive mining process. Because of the cyclical nature of the fluctuations in the overall domestic price deflator during the simulation, the behavior of income distribution shows similar fluctuations to those of the price deflator. For 7 out of the 12 years in the simulation period, there is an improvement in the share of wage income. The actual changes in the distribution of income would depend on the fluctuations in the real wage and the elasticity of the labor-demand function. One caveat applicable to any conclusions from this simulation refers to the foreign-exchange policy toward the Gran Mineria. This analysis assumes that no modifications are made with respect to the exchange rate; however, in a situation in which there occurs some significant change in production activity by the Gran Mineria, the government could reinforce its participation in the benefits by adjusting the Gran Mineria exchange rate. Such a move would affect the domestic costs of production relative to the foreign costs and lead to a substitution between foreign and domestic factors of production.

Nominal nonwage income for the Gran Mineria increases by 15.48 percent in the first year; this means an increase in government revenues through direct corporate taxes. In addition, the increase in wages is reflected in greater income tax revenues, and the higher overall volume of production increases revenues from indirect taxes. These linkages to the public sector are traced to an increase in the real current-account savings in all but two of the years in the simulation. One of the important uses of these savings is public investment, as reflected in the increase in the first year of .45 percent in total gross domestic investment, which reaches a peak after 5 years with an increase of 6.46 percent.

Noncopper Sectors

The public utilities and the tertiary sector serve as direct inputs into copper production. The former consists basically of electricity that is needed for electrolytic refining of the copper ores. The latter provides various services including transportation. The average percentage increases for these two sectors for the simulation period are 11.04 for public utilities and 3.08 for tertiary sector. The increase in production by the other sectors represents the indirect linkages of copper with the rest of the economy. Total value added increased on average by 2.95 percent. This had an impact on noncopper employment, which increased on average by 1.87 percent, or 50 thousand persons. This response shows the importance of the linkages between the copper and the noncopper sectors. On the expenditure side, noncopper investment increases as greater government revenues are channeled into the development of infrastructure. This generates more employment and adds to the productive capacity of the economy.

The GDP price deflator increases by .86 percent in the first year and then falls below the control-solution value until the last 3 years, when the price index begins an upward trend. The fluctuations in prices follow the same pattern as the money supply, even though the rate of capacity utilization increases when prices are falling. The net effect of an increase in copper production on the price level will depend on two factors: the change in international reserves, and the current-account savings by the public sector. Since it is assumed that all the increase in production is exported, in the first year the balance-of-payments surplus improves by US$39.1 million. The extent of the improvement in the balance of payments will depend on the realized export price, so that for some years the increase is much greater than for others. Thus the higher export earnings are realized as an increase in the level of international reserves. However, the increase in tax revenues from copper production as well as other sectors affected by the expansion in copper activity results in a real current-account surplus of the government with respect to the control solution. The reduced financial needs of the public sector lead to a reduction in nominal credit from the Central Bank on average of 10.8 percent per year. Whether the increase in international reserves can offset the decrease in credit to the public sector will determine how the monetary base will change and, hence, the price level.[2] However, the results of this simulation are a clear indication of the significant impacts of the copper sector on employment, income, and growth of the Chilean economy. Also affected are imports of goods and services; however, the government can offset this increase by promoting substitution of domestic for foreign inputs in copper production.

Increase in the LME Price for Copper

The purpose of this simulation is to trace the effects on the economy of having an international price of copper that is 10 percent above each historical value be tween 1960 and 1976. In effect, this takes the time path of the LME price and shifts the curve up vertically by 10 percent, with the "new" curve parallel to the original curve composed of the historical values. By means of this analysis it is possible to measure the price effect on the copper sector and the rest of the economy, just as the previous simulation considered the quantity effect from a 10 percent increase in production. By studying in detail the results of this simulation and comparing them with the one in which copper production is increased, some insights should emerge that would be relevant in the formulation of national copper policies. One of the basic conclusions of this comparison to be discussed later in this section is that an increase in production, with price variability held constant, will have a greater impact on GDP than the price increase.

Copper Sector

The initial change in the LME price does not result in an equal change in the realized export price for copper. The latter is determined endogenously in the model as a function of the LME price, and the coefficients of the equation measure the average price markdown applicable to Chilean exports of copper. The results of the simulation are presented in table 5-2. During the first year there is an improvement in the balance of payments, causing higher prices and a slight increase in GDP.

The impact of a 10 percent increase in the LME price leads to an average increase in value added by the Gran Mineria of 3.6 percent per year between 1965 and 1976. There is very little variability of value added around its period average except in 1975, when the increase in value added is 5.4 percent. These increases are determined through changes in copper output as it responds to higher relative prices and induced investment. Because of the low supply elasticity, copper output increases on average by about 1 percent per year. The difference between production and value added is a higher nominal price for copper relative to the unit costs for domestic and imported inputs. The variations in unit costs for imported inputs result from variations in the exchange rate.

Given a constant percentage change in the nominal price of copper, the effect on the relative price, the export price divided by unit costs, will depend on the cyclical response of production costs. The latter will fluctuate with nominal wages, the exchange rate, and costs of material inputs. Each of these variables will, in turn, respond to events in the economy set in motion by the original increase in the price of copper. During the first year, the increase in the relative price of copper is accounted for by the increase in the nominal price. Then as costs altered by domestic activity as well as copper-production decisions offset the increase in export price, the percentage change in the relative price diminishes during the subsequent 5 years. The only year that records a drop in the relative price with respect to the control-solution value is 1972.

The price-induced expansion in copper value added favors nonwage relative

Table 5-2
Copper Sector, A Sustained 10-Percent Increase in the LME Price of Copper
(percent deviation from control solution)

	1965	1966	1967	1968	1969	1970	1971	1972	1973	1974	1975	1976
Gran Mineria												
Value added (1965 pesos)	3.21	3.82	2.92	2.76	3.55	3.05	4.48	2.89	3.72	4.14	5.44	3.17
Export price to cost index	10.34	7.41	6.45	6.29	5.49	4.85	10.27	-12.75	10.36	14.74	7.49	7.04
Employment (000 persons)	.42	1.02	1.05	.75	.32	-.41	-1.33	-2.21	-3.74	-3.61	-3.71	-4.41
Wage bill (current pesos)	3.80	6.20	1.06	.84	-7.49	-18.60	-7.01	42.58	-8.51	17.83	15.07	7.35
Nonwage income (current pesos)	25.31	20.67	14.02	10.98	.40	-14.63	1014.3	-22.67	19.59	48.89	48.53	23.07
Production (constant 1965 pesos)												
Gross domestic product	.19	1.56	2.50	2.38	5.29	6.88	3.17	1.47	.24	-1.40	1.95	-1.01
Agriculture	0.0	-.20	.05	.33	.24	.90	1.56	.85	1.15	.97	.02	.35
Secondary sector	.42	2.11	3.13	3.01	6.91	8.82	3.89	1.79	-.09	-1.98	2.67	-1.65
Tertiary sector	.04	1.46	2.46	2.28	4.91	6.45	2.88	1.32	.37	-1.35	1.83	-.82
Demand (constant 1965)												
Private consumption	.51	2.26	2.59	2.32	5.71	6.94	4.29	2.70	1.53	-1.13	2.25	-1.04
Public consumption	-.07	.68	1.46	1.43	3.05	5.25	2.51	1.67	1.04	-.19	.51	.47
Non-Gran Mineria investment	-.48	.80	1.64	2.14	6.55	7.82	2.86	.33	-3.81	-8.87	-4.65	-9.31
Imports of goods and services	.12	2.73	4.16	3.65	7.03	7.97	3.67	1.65	.08	-1.61	2.33	-1.92
Rate of capacity utilization	-.24	1.45	2.13	1.68	5.47	7.00	1.85	-.04	-.64	-3.25	.92	-1.97
Income and employment												
Per capita net of tax wages and salaries	.36	1.64	2.81	3.12	6.02	6.84	3.39	3.05	-.69	-11.75	5.14	-1.23
Per capita net of tax nonwage income	1.74	3.07	4.01	4.47	9.00	10.06	5.05	1.70	2.29	2.71	-.60	-1.24
Employment, non-Gran Mineria	-.17	.55	.91	.90	3.04	4.48	2.72	1.53	1.17	.25	1.02	-.61

Public Sector

Current revenues (current pesos)	3.95	3.95	.25	.33	-4.22	-12.46	-4.69	-.10	-.80	19.22	19.66	7.86
Current expenditures (current pesos)	2.00	1.02	-1.21	-.52	-4.89	-9.96	-.80	-.01	0.0	12.02	10.08	4.11
Current savings (in 1965 pesos)	8.60	10.53	14.30	6.73	10.86	6.17	-127.27	2.85	4.39	75.79	84.39	15.17
Flow of credit to public sector (in 1955 pesos)	-19.99	-41.29	-34.84	-25.78	-66.58	-28.55	12.94	.72	.74	-8.41	-10.45	-10.00

Money and prices

Balance-of-payments surplus or deficit (mill. US$)	42.73	27.86	23.92	14.95	3.76	-14.51	5.26	12.10	71.14	82.45	46.67	23.46
Money supply (MI)	6.80	-.71	-9.57	-7.68	-23.01	-33.82	-5.42	-2.22	-2.23	65.18	16.72	14.33
GDP price deflator (1965 = 100)	4.37	1.62	-3.96	-2.67	-11.94	-23.00	-4.17	-1.95	-1.01	24.73	15.53	6.68
Investment (1965 pesos)	0.0	37.53	38.41	46.21	38.91	39.27	37.26	35.11	36.24	38.39	35.68	35.00

to wage income. The shift in composition is derived from a substitution between the factors of production: capital and labor. Employment is increased during the first 5 years as the copper companies demand more labor at the lower real-labor costs. In addition to more employment, the higher relative price brings about more investment by the Gran Mineria. The buildup of the capital stock contributes to the substitution of capital for labor as the companies switch to the less costly capital-intensive technology. As a consequence of this factor substitution, employment declines for the remaining 7 years of the simulation.

Income distribution in the copper sector is more favorable to nonwage income as a result of the higher price of copper. The average share of wage out of total income is 34.7 percent for the price increase and 37.1 percent for the control solution. With the exception of 1972, the share of wage income is below the corresponding values in the control simulation. The unfavorable effects of capital substitution on demand for labor contributes to the deterioration in its relative share. During the first 2 years, the wage bill increases by 3.8 and 6.2 percent, while nonwage income jumps by 25.3 and 20.7 percent, respectively.

Noncopper Sectors

Production in the noncopper sectors responds to the price increase, although with less stimulus than in the case of a 10 percent increase in copper production. Initially, GDP increases by 0.2 percent, which originates mostly in the secondary sector. Within the secondary sector, value added by the Gran Mineria accounts for almost the full increase, accompanied by some response from public utilities (with an increase of 0.7 percent). To some extent, the lack of dynamism in the economy is due to the inflationary impact of higher copper prices. Afterwards GDP begins to pick up and the secondary sector outpaces the rest of the economy as value added from the Gran Mineria maintains a steady growth.

The response by the final demand components offers some insights as to the nature of the fluctuations in production. Demand pressures during the first year have diminished relative to the control simulation, as can be observed from the decline in the rate of capacity utilization of 0.24 percent. After the first year, non-Gran Mineria investment reacts favorably to the increase in domestic activity. The income generated in the copper and noncopper sectors through the direct as well as indirect effects of higher prices results in more private consumption. The improvement in government revenues is allocated toward an expansion of government-consumption expenditures. The time path for the aggregate demand and production variables shows a smooth acceleration that peaks after 6 years and then returns to the path of the control solution.

While there is some overall redistribution of income, the degree of change is not very significant.[3] The share of wage out of total income in this simulation is 48.9 percent, and for the control simulation it is 49.3 percent. Real net of tax per capita nonwage income increases by more than the corresponding wage component in 9 out of 12 years.

Although the rate of capacity utilization is lower than the control-solution value in the first year, the GDP price deflator is 4.4 percent above the control solution. The cause for the higher prices is attributed to an improvement in the balance-of-payments surplus of 42.7 percent. In spite of a reduction in the real flow of credit to the public sector, there is an expansion of the monetary base as a result of the large inflow of international reserves. The increase in liquidity pushes prices upward. Again, in the second year, the same process is revealed: a higher level of real current-account savings by the public sector results in a reduction of the real flow of credit from the Central Bank. This reduction in the monetary base is offset by an increase in the level of international reserves derived from an improvement in the balance-of-payments account of 27.9 percent; the net impact on the GDP price deflator is 1.6 percent. The improvement in the real current-account savings by the government during the succeeding 4 years and corresponding net decrease in the real flow of credit exemplifies the importance of the public-sector budget in maintaining price stability. The increase in public savings lowers the GDP deflator to below the control-solution values between 1967 and 1973, with the exception of 1971, when savings declined significantly. In addition, the improvement in the country's balance of payments brought about by increased export revenues provides the government with the opportunity of accelerating the retirement of its burdensome foreign debt. The price impact on the balance of payments does create more inflation, as demonstrated in the first 2 years; however, government policy could deter the growth in domestic prices by sterilizing the excess revenues and drawing on them during the "bad" years.

A Comparison of the Price and the Output Simulations

A comparison of the time paths for the copper price and copper output simulations provides some interesting insights into the differences in the sensitivity of the economy to variations in prices and output. These, in turn, might furnish useful policy recommendations for stabilizing the fluctuations and developing long-term strategies for the country's copper sector. As an initial point of reference, it is useful to compare the increases in export revenues from the two simulations. The additional copper-export revenues generated from 1965 through 1976 by a 10 percent increase in the LME price are US$ 1,051.7 million, of which US$782.3 million is attributable to the Gran Mineria and US$269.4 million to the medium- and small-scale mines. The 10 percent increase in production by the Gran Mineria generates an increase of US$892.51 million. The price increase clearly contributes more foreign-exchange earnings than production, with some of the increase accruing to the medium- and small-scale mines.[4]

One of the more important differences is shown by the elasticity of GDP with respect to the price and production increases. The elasticity is calculated as the average annual percentage increase in GDP divided by the 10 percent increase in price and production. For a 10 percent increase in production, the elasticity is .295, and for a 10 percent increase in LMF price, it is .194. The relatively greater sensitivity to a production increase is due ⟩ the direct as well as indirect linkages of the Gran Mineria to other production ˊectors in the economy through input and employment requirements. In the price simulation, potential increases in noncopper production are checked by the negative impact of a higher GDP price deflator on aggregate demand.

The impact on the GDP price deflator is more favorable in the case of an increase in production. The mean percentage change of the deflator with respect to the control-solution values is -3.71 for an increase in production and +.35 for an increase in the price of copper. However, the potential for stabilization of domestic prices is greater in the latter simulation. The average percentage increase in the real current-account savings of the public sector is higher with an increase in the price of copper. This means that fewer resources were needed to finance the government deficit because the increase in the export price led to higher copper tax revenues. If this decrease in financing is accompanied by policies to offset the inflationary impact of a balance-of-payments surplus, the desirability for either of the two situations might differ from the present results. One final comment on this comparison is that the price increase gives rise to additional benefits derived from an increase in production by the medium- and small-scale mines as a result of higher prices for both types of producers.

Increase in Investment by the Copper Sector

Since the nationalization of the Gran Mineria in 1971, public policy has acquired a crucial role in the determination of long-run prospects for the Chilean economy. The issue of public investment has taken on an added dimension in the determination of the economy's production-possibility frontier. Thus the Chilean authorities are faced with the decision between the copper and the non-copper sectors. The rationale for determining whether an expansion in education or other infrastructure would be more beneficial than an expansion in copper productive capacity should be dictated in accordance with the expected benefits. By performing a simulation in which investment in the Gran Mineria is increased, it is possible to trace the multiplier effects and compare the results with those obtained for government investment in the noncopper sectors.

The proposed investment project would be in the form of machinery and equipment as well as basic construction to expand the productive capacity of the copper mines. The price tag is assumed to be US$250 million, and the project would be completed in 2 years. In addition, it is assumed that there is sufficient foreign financing in the first 3 years to cover the costs of the imported machinery and equipment as well as other material inputs. The implied financing for those 3 years amounts to US$125, US$155, and US$50 million, respectively, and is in the form of long-term credit. The results of this simulation are presented in table 5-3.

The investment in the copper sector stimulates aggregate demand in the rest of the economy. Real imports of goods and services expand by 15.1 and 14.6 percent in the first 2 years, with most of the increase represented by machinery and equipment for the Gran Mineria. Private-consumption expenditures are up 6.5 percent during the first year and are maintained above the control-solution values with a peak occurring during the sixth year of the simulation. Gross domestic non-Gran Mineria investment is also significantly affected by the investment in the Gran Mineria; however, it falls below the control-solution values during the last 4 years. Total value added increases by 7.72 percent in the first year, with a peak in the sixth year; and on average it is 6.3 percent above the control solution.

Multiplier Effect on Income and Production. The overall effects on the Chilean economy from investment in the copper industry can be measured by means of the impact multiplier as well as the dynamic multiplier. During the period of this simulation (1965–1976), the path of GDP of the "disturbed" solution is maintained above the control-solution path. The impact multiplier is calculated as follows:

$$\frac{\Delta GDPR_{65}}{\Delta IBGMR_{65}} = \frac{1.44}{.40} = 3.6$$

As can be seen from table 5-3, the GDP shows a still significant increase in the last year with respect to the control-solution values, and this expansion could persist into subsequent years, sustaining the multiplier effect. The following discussion of the behavior of the principal variables may suggest some explanations for the disparity between the copper and noncopper investment effects. Yet, in spite of the differences, investment in copper shows substantial effects on the rest of the economy.

In the production side, construction activity shows marked increases of 15.3 and 16.3 percent, respectively, in response to the investment demand. The growth in manufacturing value added is equally strong at 14.5 and 16.3 percent, respectively. The expansion of the noncopper capital stock makes possible a significant change in agricultural value added. In addition to the stimulative effect of investment, the subsequent increases in copper production explain the sustained impact on GDP.

Table 5-3

Copper Sector, An Increase in Investment by Gran Mineria
(percent deviation from control solution)

	1965	1966	1967	1968	1969	1970	1971	1972	1973	1974	1975	1976
Gran Mineria												
Value added (1965 pesos)	.22	.35	11.76	16.74	18.73	18.75	18.90	17.13	18.53	16.55	16.51	15.50
Export price to cost index	-1.36	-.39	7.64	6.40	4.08	3.36	7.56	-15.17	-1.96	-.09	.67	.05
Employment (000 persons)	-0.04	-1.56	-.71	1.25	3.26	5.05	7.06	8.18	8.20	9.04	9.93	10.75
Wage bill (current pesos)	-6.76	-16.09	-17.41	-11.21	-13.52	-16.47	-.43	49.42	-1.25	11.18	12.43	19.32
Nonwage income (current pesos)	-10.34	-18.18	-3.04	2.99	-3.92	-9.72	842.8	-23.49	18.14	15.92	20.76	23.92
Production (constant 1965 pesos)												
Gross domestic product	7.72	9.13	7.67	9.17	9.27	10.06	6.62	5.24	4.10	2.27	2.40	1.96
Agriculture	0.0	.75	1.37	1.68	2.06	2.82	3.28	3.05	3.28	2.84	2.43	2.11
Secondary sector	10.61	12.03	9.70	11.46	11.07	12.08	7.51	5.84	4.20	1.96	1.97	1.51
Tertiary sector	6.93	8.41	7.24	8.79	9.15	9.79	6.52	5.10	4.14	2.41	2.69	2.26
Demand (constant 1965)												
Private consumption	6.51	8.87	7.52	8.77	10.18	10.30	6.61	5.20	4.54	1.96	1.35	.63
Public consumption	2.47	4.60	4.87	4.88	5.99	7.07	4.01	3.16	2.41	1.65	1.19	.75
Non-Gran Mineria investment	8.15	8.54	10.72	11.14	7.79	8.23	2.04	.25	-3.81	-7.45	-9.81	-11.16
Imports of goods and services	15.06	14.57	8.38	8.84	8.08	8.04	4.72	2.71	1.90	.33	-.08	-2.13
Rate of capacity utilization	8.32	9.46	6.69	7.29	7.42	7.90	2.97	1.70	1.23	-.59	-.28	-.78
Income and employment												
Per capita net of tax wages and salaries	9.96	10.59	8.00	9.85	9.07	9.88	6.97	6.75	3.43	.47	1.91	.94
Per capita net of tax nonwage income	9.66	12.05	9.67	12.00	13.19	13.29	6.80	3.94	6.36	3.01	1.74	1.96
Employment, non-Gran Mineria	4.06	5.52	5.05	5.99	6.85	7.74	5.82	4.95	4.40	3.06	2.58	1.90

Public Sector

Current revenues (current pesos)	-.30	-6.07	-9.59	-7.01	-10.58	-13.83	-4.62	-.27	-2.51	.17	1.57	6.64
Current expenditures (current pesos)	-2.99	-6.06	-6.54	-5.48	-8.69	-10.34	-1.41	-.98	-1.80	.16	.56	3.84
Current savings (in 1965 pesos)	23.23	12.69	-5.12	1.92	6.64	4.08	-104.55	1.99	-7.46	1.32	11.68	11.48
Flow of credit to public sector (in 1965 pesos)	-51.31	-33.80	22.45	-2.28	-31.53	-14.63	10.89	-.40	1.28	-.08	-1.15	-7.73

Money and prices

Balance-of-payments surplus or deficit (mill. US$)	5.06	.53	4.05	-.46	10.35	9.74	10.82	22.16	82.53	67.68	-120.05	29.52
Money supply (MI)	-24.08	-33.76	-28.62	-26.22	-31.64	-34.57	-8.79	6.58	-5.85	9.04	4.65	13.08
GDP price deflator (1965 = 100)	-8.60	-16.71	-17.69	-15.04	-20.97	-24.99	-6.91	-5.24	-5.86	-1.07	-.28	5.72

Copper Sector

Within the copper sector there is an increase in production in the third year, when the installation of the machinery and equipment is completed. The initial increase in value added is 11.76 percent and rises to 16.74 percent in the fourth year. During the 10 years that output responds to the additional capacity, value added responds to the additional capacity. In terms of physical quantity, the output from the Gran Mineria increases on average by 108.9 thousand metric tons per year.

The copper sector does make significant gains in productivity, and these are reflected in more favorable relative prices for five consecutive years and a slight decrease in employment from the substitution of capital. However, this unfavorable employment effect is eliminated beginning with the third year, when production requirements lead to significant increases in employment. The change in distribution of income between wage and nonwage income for the copper industry is slightly in favor of the nonwage component. The average share of wage out of total income for the period of the simulation is 36.2 percent compared with 37.1 percent in the control solution. This change occurs as a result of the increase in investment, which generates higher profits through improvements in the productivity of capital.

Noncopper Sectors

Overall income distribution in the economy shows a small change in favor of nonwage income, reflecting the redistribution within the copper sector. The average share of wage out of total income for the economy declines to 48.9 percent from 49.3 percent in the control solution. Even as the share of wage income deteriorates, there are significant increases in noncopper employment in response to the higher level of domestic activity. The combination of the large expansion in copper investment and the demand-induced increase in noncopper investment brings about a rise in total labor productivity during the first 8 years. However, the growth of real per capita wage income does not match that for the nonwage component.

The impact on monetary variables is realized through an improvement in the balance of payments accompanied by greater price stability. For 11 out of the 12 years of the simulation, the GDP price deflator is significantly below the control-solution values. This is achieved through a more favorable position for the current account of the public sector. The income generated in the copper and noncopper sectors of the economy leads to higher levels of government revenues, and since expenditures do not adjust immediately, the real current-account public savings improve by 23.2 and 12.7 percent, respectively, during the first 2 years. The lower government-budget deficits mean that not as much financing from the Central Bank will be forthcoming, moderating the inflationary impacts of the government-sector budget.

A Comparison of Copper and Noncopper Investment

There are some similarities as well as contrasts between this simulation and the one dealing with public-sector investment. In both cases there is an immediate increase in production capacity. The cyclical movements in output are very similar. The differences are most pronounced within the money and prices sectors. In the previous noncopper investment case, there is a strong deterioration in the balance of payments, with the exception of the last 2 years. This is the result of high imported-inputs requirements in view of greater domestic activity, without an offsetting increase in the exports of goods and services. With copper investment there is a noticeable improvement in the balance of payments in 10 out of the 12 years as output of copper responds to the larger stock of capital. The path of the GDP price deflator is lower in the case of copper investment in spite of a higher level of international liquidity. The differences in the price deflator are explained by the more extensive use of real flow of credit to the public sector to finance the noncopper government-investment program.

One quantitative index that could be used to compare the impact from the two types of investments over the period of simulation would be the increase in the average value of GDP divided by the initial investment expenditures. The corresponding values of this ratio for copper and noncopper investments are 2.18 and 2.09. Even though the investment assumptions differ in some respects, the values of the dynamic-impact index appear to be reasonably close, so that neither one holds a clear advantage in generating more employment. However, an expansion of investment in the copper sector results in a much more favorable balance-of-payments situation.

Gran Mineria Exchange-Rate Policy

This section deals with one of the more controversial government copper policies: the specific exchange rate applicable to the purchase of domestic currency by the Gran Mineria. Whenever the copper companies need to make payments to Chilean employees, to domestic suppliers of material inputs, and to the government for tax purposes, they must purchase from the Central Bank the necessary amount of pesos with the dollar proceeds of their exports at an exchange rate specified for these transactions. This exchange rate has been used as a policy instrument to augment the availability of foreign exchange to the public sector. Thus, since the government maintains this rate below the official one, companies must pay more out of their earnings to purchase domestic inputs.

In the analysis of government policies presented in chapter 2, references were made to the exchange-rate policy and its sporadic use. Between 1960 and 1976, the only significant use of an overvalued Gran Mineria exchange rate was during the Allende period. This policy continued into the first year of the Pinochet administration.[5] Although the differential exchange rate provides the Central Bank with "easy dollars," its critics claim that it distorts production and leads to a substitution of foreign for domestic inputs. The latter effect contradicts the original purpose of the policy, which is to increase Chile's participation in the copper industry. Since the nationalization of the Gran Mineria, the exchange-rate policy became counterproductive in view of the fact that the government would receive all foreign-exchange earnings from copper exports and these in turn would be allocated to development projects as well as other government expenditures. The only incentives for using an overvalued exchange rate might be political, in that the value of foreign-exchange differentials would be accruing to the Central Bank and this would make it easier for the central government to appropriate more funds without undergoing lengthy debates in Congress to approve increased outlays.

This simulation quantifies some of the arguments expressed for or against the differential exchange rate. A value of the Gran Mineria exchange rate that is 10 percent below the control-solution value for every year is used as the representative "repressive" policy scenario. The results are shown in table 5–4, which covers the period 1965 through 1973, after which the "disturbed" solution has been allowed to return to the path of the original control solution. Even though the differences between the two solutions are small, they point out some interesting and in some cases expected conclusions.

Copper Sector

The immediate impact of the change in the exchange rate is an increase in production costs. The export price to the cost-of-production index, which is measured in U.S. dollars, drops by 9.4 percent. This ratio is a principal explanatory variable in the copper-production equation. The size of the impact on the relative price is explained by the high proportion of domestic inputs (labor and materials) that enter into the production process. During the period 1965 to 1973, the average share of domestic inputs out of the total in U.S. dollars was 87.0 percent, with the corresponding share for the pre-Allende period being 84.5 percent, and for the Allende period, 92.8 percent. On average, the relative price index decreased by 8.6 percent.

As a result of greater costs, copper production and value added decrease by an average of 1.4 and 1.0 percent per year, respectively. The very slight increases in value added for the first and second years reflect a revaluation of the official exchange rate in response to the differential between the domestic and external

Table 5-4
Copper Sector, A Sustained 10-Percent Decrease in the Gran Mineria Exchange Rate
(*percent deviation from control solution*)

	1965	1966	1967	1968	1969	1970	1971	1972	1973
Gran Mineria									
Value added (1965 pesos)	.55	.03	-2.47	-2.37	-.20	-.04	-1.91	-.90	-1.39
Export price to cost index	-9.38	-10.32	-9.97	-8.96	-9.38	-9.24	-9.69	9.49	-20.13
Employment (000 persons)	0.0	-.11	-.73	-1.23	-1.04	-.93	-1.20	-.98	-.39
Wage bill (current pesos)	-.24	-.02	-.76	-2.07	-1.47	-1.34	-1.13	-37.50	5.86
Nonwage income (current pesos)	.22	-.39	-5.04	-3.36	-.04	.24	-93.72	35.00	-2.30
Production (constant 1965 pesos)									
Gross domestic product	-.07	-.03	-.23	-.26	-.20	.08	-.41	-.54	-.93
Agriculture	0.0	0.0	-.04	-.07	-.11	-.16	-.19	-.31	-.35
Secondary sector	-.06	-.07	-.36	-.39	-.29	.04	-.57	-.69	-1.28
Tertiary sector	-.10	0.0	-.17	-.19	-.15	.16	-.33	-.45	-.74
Demand (constant 1965)									
Private consumption	.67	.72	.08	-.11	.57	.01	-.33	-.33	-.78
Public consumption	.15	.13	.03	-.02	.06	.07	-.05	-.13	-.21
Non-Gran Mineria investment	-.22	-.47	-.61	-.79	-.83	-1.11	-.69	-.87	-1.70
Imports of goods and services	.05	-.02	-.05	-.23	-.18	-.19	-.25	-.22	-.70
Rate of capacity utilization	.27	.34	.38	.30	.57	.03	.28	.26	-.29
Income and employment									
Per capita net of tax wages and salaries	-.12	-.23	.22	-.08	-.18	-.05	-.33	-1.59	-.88
Per capita net of tax nonwage income	.36	-.53	-2.11	-1.50	-.32	.96	-2.15	1.11	-1.31
Employment, non-Gran Mineria	.15	.12	.02	-.07	.09	-.09	.06	.01	-.30
Public Sector									
Current revenues (current pesos)	-.33	-.29	-.56	-.55	-.27	-.10	-.61	-.72	.30
Current expenditures (current pesos)	-.15	-.08	-.24	-.24	-.08	-.07	-.19	-.09	.55
Current savings (in 1965 pesos)	-.89	-1.26	-1.88	-1.83	-.97	.36	-13.70	5.17	.17
Flow of credit to public sector (in 1965 pesos)	2.06	3.62	6.39	7.22	6.48	-2.03	1.37	1.24	.02
Money and prices									
Balance-of-payments surplus or deficit (mill. US$)	-4.77	-5.60	-8.71	-4.26	-4.91	-7.39	-1.46	2.10	.27
Money supply (M1)	.51	1.02	1.87	.99	1.45	-.10	.93	1.64	2.79
GDP price deflator (1965 = 100)	-.32	.06	.05	.01	.18	-.52	.03	.14	1.40

inflation rates. The cutback in production leads to a reduction in employment. There are two reasons why the demand for labor is less with an overvalued exchange rate: first, the real cost of labor is higher; and second, the drop in output requires less of the variable factors. Both reasons contribute to an average decrease in employment of 0.83 percent per year, which implies an average elasticity of .08 with respect to exchange-rate changes. The combination of a reduction in output and the factor substitution in favor of capital induces a redistribution of income in the copper sector. The average share of wage out of total income is decreased from 40.3 percent in the control solution to 38.8 percent with the overvalued exchange rate for the Gran Mineria.

Noncopper Sectors

In the noncopper sectors, GDP is lower in 8 of the 9 years. During the first 2 years, the variation in GDP is not significant; then, as value added in the copper sector begins to decline, concomitant with a drop in the services sector, total domestic output is decreased by 0.23 percent in the third year, maintaining a somewhat steady rate of decrease throughout the period. To some extent, the contraction in domestic activity is brought about by the lower production rate in the Gran Mineria, which then affects the other sectors via the direct and indirect linkages to employment and income.

The slight decline in domestic activity is reflected in lower imports of goods and services, which are below the control solution from the second through the last year of the simulation. Gross domestic investment is also adversely affected, more so than other aggregate demand components. Because of the diminished availability of productive capacity and imported inputs, the rate of capacity utilization is maintained above the control-solution values. The path for the income variables is more apparent in the case of real per capita net of tax wage income, which decreases during the first 2 years, shows a small upturn during the third, and remains below the control-solution values for the remaining years. The nonwage component does not display a consistent pattern. These factors are exemplified through the variable-path behavior of consumption.

A principal area of concern is the sector of money and prices. The deterioration in the balance of payments during the first 7 years is explained by a reduction of copper production, which leads to a greater decline of export earnings than the reduction in the value of imports. The GDP price deflator is 0.32 percent lower in the first year as a result of a greater external deficit, even though the real flow of credit to the public sector is increased by only 2.1 percent. Subsequently, the application of greater resources to finance the public-sector deficits induces a positive change in the price deflator.

The policy implications of this simulation are that the use of an overvalued exchange rate for the copper industry is worse than using the official commercial bank exchange rate. By raising the costs of production for the Gran Mineria, it penalizes production, leading to reductions in export earnings and public-sector revenues. The consolidation of the copper-sector response with the ensuing repercussions on domestic activity yield a value of GDP slightly below the control-solution value. At present, the desirability of such a policy is questionable given that all foreign-exchange earnings are recuperated by the government.

Conclusions

Some of the results of the copper-sector simulations are interesting in terms of their policy implications. The cases presented in this chapter do not exhaust all the possibilities, since this would involve dozens of simulations; however, these four cases can be used as a basis for answering other policy questions without the need for more simulations. The next chapter will deal with the effects of copper-price variability, and chapter 7 will outline feasible policy alternatives for the management of the copper industry that would meet the overall development objectives of the Chilean economy.

Notes

1. See chapter 4 for an explanation of the multiplier formula.

2. During the first year of the simulation, the balance-of-payments effect on the monetary base is stronger than the change in credit from the Central Bank to the public sector.

3. This represents the total share of wage income in the copper as well as noncopper sectors.

4. Actually the difference in export earnings is attributed to the application of higher prices to both the Gran Mineria and the medium- and small-scale mines, while the production increase refers to the Gran Mineria only.

5. After October 1974, the exchange rate applicable to all Gran Mineria transactions was the official commercial bank rate.

6

Copper-Market Fluctuations and Domestic Activity

One of the underlying causes of the international movements to establish commodity price-stabilization schemes is the notion that export instability affects domestic income, investment, and long-term growth. The countries that are particularly keen on implementing international commodity agreements are those for which the share of a primary commodity out of total export proceeds is exceptionally high.[1] The variability in earnings from primary commodity exports would result in large fluctuations in the balance-of-payments account with consequent impacts on the rest of the economy. These impacts could be in the form of increased fluctuations of domestic variables (as measured by the standard deviation of the variable) or an alteration of the economy's growth path. The aim of this chapter is to present a brief analysis of the principal issues of the export-instability literature and to test the validity of the variability hypothesis in the case of Chile.

The methodology for testing the variability hypothesis will consist of two model simulations during the historical period 1965–1976, one with a smooth and the other with a fluctuating price of copper. By comparing the time path of the important variables for each case, it is possible to identify the specific areas most affected by price instability. These areas would include the balance-of-payments account, the public-sector budget, and the overall price level. The outcomes of the simulation will indicate to what extent the problem of export instability has affected the Chilean economy. The basic question is whether export price variability has really been a problem for Chile. The results of the simulation also will indicate what policy measures may be adopted to offset some of the negative consequences of price instability on the copper industry. In the following chapter, a more detailed analysis of these options will be considered.

The nationalization of the copper sector has established a new role for the government as the owner of a very important industry, and because of this new dimension in the ownership of the copper mines, the government needs to review its strategies in dealing with the potentially disruptive effects from fluctuations in export earnings.

The debate on export instability has centered on the issue of whether export instability has a negative, a positive, or no effect on the economic development of less-developed countries. This actually involves two questions: What is the impact in the short run? and What is the long-run impact on growth? The first deals with the immediate effects through a shortfall in export earnings on imports and total domestic output. From year to year, these short-run responses may cancel out through the ups and downs in the price of exports. While in the long run, the accumulation of these short-run effects may result in differing growth trends for the economy. In addition, the existence of a secular price trend may invariably determine the growth path of the economy regardless of the sequence or intensity of individual short-term fluctuations. The analysis of export instability in this chapter will focus on the issue of price variability and not on the consequences of secular trends.

The empirical discussion on export instability has drawn on two types of analyses: cross-sectional analysis and time-series analysis. The former involves the collection of relatively homogeneous group of countries, and the latter deals with one particular country over time. Most of the published studies conclude that the case-study method may be a more fruitful approach. In fact, Maizels (1968) and Ragarajan and Sundararajan (1976) assert the need to analyze the links on a case-by-case basis with sufficient disaggregation to trace the complex multiplier processes. The use of cross-sectional data to estimate the correlation between export instability and economic growth implicitly assumes an identical economic structure across countries. A simple correlation between these two variables does not reveal the intricate mechanism by which the effects of export fluctuations are transmitted to the rest of the economy, and as MacBean (1966) and others have shown, the institutions as well as economic structures differ significantly across countries.

A Priori Hypotheses

There are several hypotheses concerning the relationship between export instability and domestic activity.[2] A summary of two of these will follow; the first describes the negative effects, the other why instability will have little effect. The general line of reasoning in the first case contends that reduced earnings result in reduced import capacity—directly, through a reduction in foreign exchange, and indirectly, through increased costs of foreign borrowing as interest rates charged by foreign banks adjust to reflect the greater uncertainty.[3] The direct effect of a reduction in foreign-exchange earnings is exemplified by various restrictions on imports, which may include basic capital equipment. Gross domestic investment would be adversely affected by reductions in machinery imports. The indirect uncertainty effect would also establish an unfavorable environment for investment as businessmen question the feasibility of a project that may have to be abandoned or postponed because of shortages of imported equipment. In addition, the risks associated with instability could bring about a rise in inventory investment instead of an expansion in the productive capacity of the economy. The reductions in the quantity and the efficiency of investment would deter the long-run growth of the economy.

The reduction in export earnings would affect income in the export industry. This would curtail investment as well as consumption expenditures by that industry. The accumulation of all the preceding consequences would result in a multiplier effect on GDP whose size would be determined by the extent of the linkages between the export sector and the rest of the economy.

This hypothesis emphasizes the growth issue. Yet, it is also association with the short-run effects originating from a sudden and temporary reduction of import capacity or the increase in overall prices as brought about by a boom in export prices. The latter case arises from the excess liquidity that is created by transferring export earnings into domestic demand for goods and services. If the economy is operating at close to full capacity, the excess demand will lead to higher prices. Whereas in the long run there may be no clear indication of the effect of export instability, in the short run there are always some variables that are adversely affected to the extent described in the previous paragraphs.

The second hypothesis is based on Friedman's theory of consumption. Knudsen and Parnes (1975) assumes, as in the permanent-income hypothesis, that there are no correlations between the transitory and permanent components of either income or consumption and none between the transitory components of income and consumption and that exporters of primary commodities are able to save the increased earnings from exceptionally good years. Thus the fluctuations are mainly absorbed by the exporters, who offset the bad years with savings accumulated during the good ones. However, the success of this auto-stabilizing process is dependent on the ability of the government to avert a deficit-induced inflation during the low-earnings years. For countries such as Chile with chronic burdensome government deficits, there would be an element of domestic price variability brought about by the inflexibilities in the government budget or the political system. This second hypothesis explaining why instability will have little effect is supported by MacBean, who concludes from the cross-sectional analysis that "fluctuations in GNP appear to be quite heavily damped."[4]

Empirical Studies

In chapter 3, some econometric tests of the export-instability hypothesis were presented. These dealt with the impact of variability, as measured by the standard deviation of export earnings, on government expenditures and various import components. By including the variability term in the equations for each of the latter variables, a test was made of the negative effects of the fluctuations in export earnings. The results showed that the instability variable included in the equations for investment, public consumption, and imports was not statistically significant, and in a number of cases, the coefficient had the wrong sign. Out of fifteen coefficients for the variability index, three were significant but had the wrong sign. These were (1) imports of consumption (positive) with respect to total export earnings, (2) government consumption (positive) with respect to foreign reserves, and (3) material imports (negative) with respect to foreign reserves. This section pursues another test of the instability issue through simulations of the econometric model. While the econometric tests imply that there may not be any significant impact from expectations, the indirect effects, as they pertain to monetary and fiscal policy, may support the hypothesis of some linkage between instability and domestic activity.

The advantage of using an econometric model to test for the relation between export instability and domestic activity is that the complex economic decision-making process represented by the model permits a more complete measure of the multiplier effects described in the first general hypothesis. In addition, if the economy is heavily damped, as asserted by McBean, simulations with an econometric model are best suited for testing this type of characteristics. While a simple correlation between two variables will indicate how sensitive the one is to fluctuations in the others, this type of analysis is incomplete. A simultaneous econometric model represents a more complete analysis of the complex interactions between the affected variable and all the others in the system. For instance, if only one equation is used to measure the impact of variability, for example, the price equation, an increase in export prices would increase domestic liquidity (one of the explanatory variables) and, hence, the price level. However, if the price equation is part of a simultaneous model, the increase in prices arising from excess liquidity would reduce aggregate demand or the rate of capacity utilization (another explanatory variable). The latter would bring down the overall price level, so that the net effect from the increase in export prices would be less than the initial reaction to excess liquidity. The use of model simulations that incorporate the simultaneous offsetting or reinforcing effects will determine how damped an economy is in responding to exogenous changes that would not be possible through a partial, or single-equation, analysis.

Rangarajan and Sundararajan (1976) combined the use of a simultaneous-equation model for eleven developing and three developed countries and compared the results of the income multipliers. Their methodology consisted of estimating a very simple (six equations) Keynesian model for each of the countries, thus imposing an identical and perhaps inappropriate economic structure on each one. Using the estimated models, they calculated the impact, intermediate, and long-run multipliers to check for a consistent effect across countries of a smooth export path. In only five of the countries was there any adverse effects of export instability. The use of the simplified models may be questionable, and as they conclude, "the policy implication is that the usefulness of international schemes for stabilizing primary product prices or the export earnings of less developed countries has to be examined for each country separately on the basis of an analysis of its economic structure".[5] Their study fails to adequately incorporate into the econometric models the complexity of the multiplier process, the supply constraints, and the differences in economic structure and policies.

The case-study approach reveals more information on the nature of the adjustment process and the cyclical behavior exhibited by different economies. One good example in the application of econometric method to a case study is Adams and Roldan (1978) for Brazil. Although they evaluate the implications of production and pricing decisions in the coffee sector on the rest of the economy, their analysis does not deal specifically with the issue of instability.

An econometric analysis of export instability in the case of Chile has not been attempted previously; notwithstanding, MacBean provides a good synthesis of the data on Chile and some interesting conclusions.[6] He describes copper in Chile during the prenationalization period as a classic example of an enclave dominated by a foreign enterprise. This is an important consideration in analyzing fluctuations in export earnings, since only a portion of those earnings is returned to the domestic economy,[7] so a large share of fluctuations in export earnings is transmitted via remittances of dividends to the foreign stockholders. MacBean's argument is that in the case of Chile, the important variable which the government should respond to in stabilizing the domestic economy is the "returned value" of exports. The latter consists of payments for domestic imputs and tax liabilities, of which the principal component is the generation of tax revenues. In reference to the susceptibility of the public sector, MacBean comments that "fluctuations in revenues from copper are one cause of inflation in Chile through generating budget deficits when the value of copper exports falls below previous levels."[8] Since the government's capacity to raise other taxes or cut back on current expenditures is rather limited, this option faces some serious predicaments as a tool for offsetting export instability.

An Econometric Model Simulation of the Effect of Export Instability on Chile

The preceding discussion has brought out several hypotheses that attempt to explain why export instability has a negative, positive, or no effect on domestic activity. In chapter 3, a partial econometric test of some of the direct linkages was inconclusive. This section presents the results of the case-study approach to testing those hypotheses through the use of an econometric model.

This study will focus on the impact of price variability, not total export earnings. In the case of Chile and copper, prices have fluctuated much more than quantity. The normalized variability index for output from the Gran Mineria between 1960 and 1976 is .18, and for the LME price it is .38.[9] Because of the widespread discussion of commodity price-stabilization schemes as well as the gains from cartelization following the formation of CIPEC in 1967, the price issue is particularly relevant. The construction of a smooth price series was accomplished by means of a linear regression of actual price on a time trend. The values calculated with the equation were used as the smooth alternative to

Table 6-1
Smooth versus Variable LME Price of Copper
(percent deviation from control solution)

	1965	1966	1967	1968	1969	1970	1971	1972	1973	1974	1975	1976
Gran Mineria												
Value added (1965 pesos)	.05	-5.56	4.00	1.64	-1.41	2.13	7.58	8.00	-7.41	-15.67	10.74	3.52
Export price to cost index	-1.62	-16.21	23.52	1.69	-8.03	-.36	20.35	-7.61	-14.64	-36.90	21.18	7.73
Employment (000 persons)	0.0	-.68	-.31	.91	.63	.45	.79	1.07	-.17	-1.45	-1.90	.81
Wage bill (current pesos)	-.25	-6.56	2.74	14.53	-2.39	-3.00	3.84	76.49	2.66	-38.51	-30.17	32.19
Nonwage income (current pesos)	-2.60	-35.96	62.95	15.21	-18.49	1.71	2172.9	17.30	-31.70	-74.06	35.81	57.18
Production (constant 1965 pesos)												
Gross domestic product	0.0	-.47	-2.47	.24	-.01	.29	-.27	-.05	-.48	-.80	-7.05	-1.65
Agriculture	0.0	0.0	.30	-.91	-.29	.04	-.54	-.84	-.88	-.58	1.09	-1.28
Secondary sector	.13	-.92	-3.27	.49	-.34	.33	-.13	.07	-1.15	-1.92	-10.20	-2.46
Tertiary sector	-.10	-.21	-2.36	.25	.31	.30	-.34	-.03	.10	0.0	-6.48	-1.15
Demand (constant 1965)												
Private consumption	.65	.32	-1.41	-.23	.56	.08	-.10	.57	-1.00	.97	-6.74	-.51
Public consumption	.15	.41	-1.52	-.51	.74	.36	.02	.20	.90	.76	-2.14	-2.85
Non-Gran Mineria investment	-.22	.30	-4.67	-1.36	-2.18	-.57	-1.96	-1.70	-.96	3.30	-17.38	-9.41
Imports of goods and services	.07	.01	-4.73	1.04	1.02	-.26	-.19	1.04	1.38	.47	-11.15	-2.67
Rate of capacity utilization	.27	1.34	-2.78	-.07	1.17	.51	-.58	-.17	.29	4.52	-6.91	-1.81
Income and employment												
Per capita net of tax wages and salaries	-.25	-.42	-3.08	1.20	-.16	-1.09	.01	2.35	1.29	12.61	-29.96	-6.12
Per capita net of tax nonwage income	.48	-3.66	-2.23	.82	.04	.09	1.81	-1.92	-2.58	-6.54	6.49	3.42
Employment, non-Gran Mineria	.16	.65	-1.43	-.58	.12	-.07	-.56	-.60	-.52	.98	-1.55	.24

Public Sector

Current revenues (current pesos)	-.35	-6.81	6.50	7.45	-3.64	-1.16	2.83	5.55	.31	-38.88	-26.76	22.07
Current expenditures (current pesos)	-.15	-3.49	5.84	2.74	-3.30	.20	1.54	2.10	.30	-20.73	-10.07	17.10
Current savings (in 1965 pesos)	-.96	-14.56	-4.63	19.95	2.23	-4.19	38.09	-30.99	.10	-423.71	-203.56	6.83
Flow of credit to public sector (in 1965 pesos)	2.21	42.83	12.78	-80.52	-12.03	25.74	-4.00	-7.55	.02	44.96	22.95	-9.67

Money and prices

Balance-of-payments surplus or deficit (mill. US$)	-6.48	-59.38	139.53	17.56	-26.78	-11.42	54.46	67.63	-125.63	-166.05	369.21	24.13
Money supply (M1)	-.93	-7.69	29.62	6.16	-9.70	.40	7.40	5.48	-4.13	-58.74	26.93	57.02
GDP price deflator (1965 = 100)	-.33	-7.06	14.45	6.02	-6.75	-.12	3.55	4.94	.18	-40.97	-13.81	32.87

Table 6-2
The Actual LME Price of Copper and the Smooth Price Based on a Linear
Time Trend, 1965-1976
(U.S. ¢/lb)

	LME Price	Smooth Price
1965	58.34	57.48
1966	69.31	58.50
1967	51.08	59.53
1968	56.21	60.55
1969	66.29	61.57
1970	64.13	62.59
1971	49.26	63.62
1972	48.55	64.64
1973	80.77	65.66
1974	93.39	66.68
1975	55.99	67.71
1976	63.92	68.73

Note: The smooth price was generated by the linear regression: Price = a + bTrend; LME
price: Average = 63.10, standard deviation = 13.31; smooth price: Average = 63.10, standard deviation = 3.69.

Table 6-3
Mean Value and Standard Deviation of Principal Variables in Smooth versus
Variable Price Simulation, 1965-1973

	Variable Price[a]	Smooth Price[a]
GDPR (1965 pesos)	21.576 (1.8)	21.50 (1.06)
Inflation	16.616 (155.7)	86.86 (146.4)
Balance of payments surplus/deficit (mill. US$)	24.37 (218.5)	38.41 (188.11)
Noncopper investment (1965 pesos)	2.79 (.09)	2.75 (.09)
Exports of goods (mill. US$)	1,019.39 (168.6)	1,033.92 (150.39)
Imports of goods (mill. US$)	940.90 (228.8)	941.4 (230.0)
Gran Mineria copper production (000 M.T.)	584.90 (25.08)	584.58 (25.62)

[a]The standard deviation is shown in parentheses.

the fluctuating historical price, and as a result of the least-squares method, both series have the same mean.[10]

The period of simulation is 1965 to 1976, with the control solution, based on the actual LME prices, serving as the reference solution. The results of the simulation are presented in table 6-1. Table 6-2 shows the actual LME price and the smooth price used for the alternative solution. In addition, to compare overall differences between time paths, the arithmetic mean and sample standard deviation of the principal variables have been calculated in table 6-3. A comparison of the mean value from both simulations serves to indicate any differences attributable to copper-price variability. The sample standard deviation measures variability of each of the principal variables and is indicative of the sensitivity of domestic variables to fluctuations in export earnings. Whereas average values may show no significant differences, the variability in each case may differ; the less the value of the standard deviation, the less is the economic uncertainty facing economic decisionmakers.

Although the variable versus smooth price simulation refers to the 1965–1976 period, this may not be the appropriate period for comparison. The radical political changes make it difficult to separate out the underlying effects of export instability on the Chilean economy. This complication is quite apparent in the 1974–1976 period, in which the repressed inflation of the Allende period leads to hyperinflation following the extensive price-liberalization schemes of the Pinochet administration. In response to the chaotic prices, the Pinochet administration applied austere fiscal and monetary policies. Because of the sensitivity of the price equation to any variations in the explanatory variables coupled with the great disparity in the rates of inflation between the 1974–1976 and 1960–1973 periods brought about through political instability, it follows that a reasonable period for evaluating the impact of price variability is 1960–1973.

A Comparison of Variable- and Smooth-Price Paths, 1965-1973

Since the purpose of this analysis is to test some of the hypotheses dealing with export instability and domestic activity, the comparison will begin with a summary of the impact on those variables prominent in the export-instability literature: GDP, prices, investment, and imports. The differences, as shown on table 6-3, are more pronounced in terms of variability (as measured by the sample standard deviation). The differences for GDP are not statistically significant, as demonstrated by the mean value for the period 1965–1973; with price variability it is 21.6, and with smooth prices it is 21.5. The corresponding sample standard deviations are the same at 1.8. Even for the 1965–1976 period, the average values for GDP are almost identical, with 21.7 and 21.5, respectively. The equality in the variability measures for GDP is the result, to a large extent, of the lack of built-in response to a change in the path of copper prices as actual fiscal and monetary policies have worked to isolate the rest of the economy from fluctuations in the world price of copper. In addition, because of the relatively small size of the copper sector, any variation in copper output has resulted in a minimal change in real output of the economy.

The average rate of inflation (measured by the percentage change in the GDP price deflator) is 89.1 percent with the variable price of copper and 86.9 percent with the smooth price. The standard deviation for the rate of inflation is, for the variable and for the smooth price, 155.7 and 146.4 percent, respectively. By comparing the results for GDP and the domestic price level, it is evident that export-price instability has had little impact on real output, but it has induced higher domestic prices as well as greater variability in the price level. This basic conclusion differs from the a priori theories that ascribe much greater importance to the negative effects on real output. The increase in domestic price variability caused by variations in the price of copper, as brought out by the simulation, will affect the distribution of income and the balance of payments as well as other sectors. The desirability of implementing economic policies to offset the negative impact of export variability on domestic activity will be discussed in the following chapter.

The decrease in variability of the GDP price deflator with the smooth copper price points to an important linkage between export instability and domestic activity. This operates through two channels. The first is the variations in the foreign reserves that affect the monetary base and, eventually, the domestic price level. Whenever the actual export price of copper is below the smooth price, there is a shortfall in export earnings. The reduction in foreign-exchange earnings will be reflected in a higher current-account deficit. This, in turn, will bring down the inflow of international reserves into the Central Bank. Since the monetary base, or high-powered money, is defined on the asset side as the sum of international reserves and Central Bank credit to the public and the private sectors, the reduction in international reserves will have a direct negative impact on the money supply (M1). This occurs through a cutback in the expansion by commercial banks of their loan portfolios. The contraction of credit availability would cool off excess demand and initiate some downward pressures on the general price level.

The second element of the linkage operates through the public-sector budget. Copper price variability would affect government savings and, hence, gross domestic investment and the money supply. If there is a sharp decrease in copper export prices, tax revenues from the copper sector would contract. This would be reflected in a reduction of government outlays. In the short run, a cutback in current expenditures of the public sector to offset the shortfall in revenues would be very troublesome politically. The item that is probably more susceptible to changes as a result of a shortfall in revenues would be capital expenditures. As explained in chapter 3, gross investment in the noncopper sectors is linked to the level of government savings in constant pesos. A decrease in public-sector revenues would result in a reduction of total investment and, hence, overall economic activity.

The connection between the public-sector budget and the money supply occurs through the Central Bank financing of the budget deficit. If the price of copper decreases while government outlays are not adjusted downward sufficiently, the overall deficit is increased. The Central Bank would have to extend additional domestic currency financing to offset the decrease in foreign-exchange revenues. The growth of high-powered money used to finance the higher deficit would expand domestic liquidity and exert upward pressures on prices. Also, with less foreign exchange available, the government would compete with the private sector for the scarce foreign currency. The latter sector could meet with import restrictions that could eventually bring down domestic economic activity. This impact of foreign-exchange availability on imports is incorporated in the consumption-goods equation (see chapter 3).

Effect of Price Variability on Domestic Variables

Gross domestic investment in the noncopper sectors is about the same in the variable as in the smooth price simulations; the average level is marginally lower with the smooth prices. The average values for the variable and the smooth copper prices are 2.79 and 2.75 million 1965 pesos, respectively, and the sample standard deviations are both .09. The value for the standard deviation of noncopper investment does not change between the two simulations, as is also the case for GDP. Imports of goods are somewhat lower with a smooth price. The infinitessimal difference between the average values for imports measured in millions of U.S. dollars can be ascribed to sporadic shortfalls in domestic activity with a smooth price, especially the reduction in demand for capital-goods imports. While imports of goods register no significant changes, the surplus on the balance of payments shows a significantly higher average value and a lower variability measure in the case of a smooth price of copper.

Effect of Price Variability on the Copper Sector

In the Gran Mineria, the impact of smooth copper prices is heavily damped because of supply inflexibility. With a very low elasticity of supply, copper producers are unable to respond in full measure to the incentives from stable prices. Surprisingly, the mean value of production is slightly lower with smooth prices— 584.58 thousand metric tons. The maximum response to the smooth-price series occurs in 1966 and in 1974, when production declines by 2.4 and 4.0 percent, respectively. In 1966, the smooth price was 16 percent lower than the actual price, and in 1974, it was 29 percent lower. Those 2 years were characterized by historically high copper prices.[11] The fluctuations in value added by the Gran Mineria are due to exchange-rate variability, fluctuations of domestic prices, and other costs of inputs relative to the price of output. Even though there are some significant year-to-year changes, the average value of value added for the period is 1.067 with variable and 1.076 with smooth prices—almost identical. For the ratio of export price to costs of production, the average value with variable prices is 101.1, and with smooth prices, 99.4. This phenomenon reflects the element of variability introduced into the costs of production, but which does not alter the average level of production. The lack of supply response to changes in the price-to-cost ratio restricts the potential for the export sector to significantly affect the rest of the economy through the income-multiplier process.

The changes in the income distribution induced by the smooth copper prices are definitely in favor of wage income. The average share of wage out of total income is 37.1 percent with the variable and 43.32 percent with the smooth copper prices. The income redistribution occurs through the costs of production. One of the principal components, the nominal wage, shows an average value with the smooth prices that is about 11 percent higher than with the variable prices, while the average number of workers employed remains virtually unchanged.

Effect of Price Variability on the Noncopper Sectors

The cyclical behavior of the noncopper sectors follows the same pattern as that shown by the copper prices. However, the ups and downs of prices in the smooth-variable price simulation obscures the cause-effect relationship since the lagged effects offset any sudden change in the direction of prices. For example, for the first 2 years of the simulation (1965–1966), the smooth price happens to be less than the actual price. Those variables which are affected by the export price would determine the course of economic activity into the following years, despite the fact that the smooth price is above the actual price during 1967–1968. This phenomenon is caused by the lagged response of a dependent variable to its explanatory variables. Since government-consumption expenditures in constant pesos adjusts slowly to its desired level (as determined by tax revenues), the negative impact of a shortfall in revenues in one year will carry over into subsequent years. In other words, full adjustment takes several years to complete. These dynamic characteristics of the model explain why the higher smooth price (relative to the actual price) during 1967–1968 does not completely offset the impact of the lower smooth price in 1965–1966. In fact, if the first year of the simulation had been one in which the smooth price was greater than the actual historical price, the results may have been somewhat different.

For this reason, it is preferable to focus on a comparison of the average values and the standard deviation of the principal variables. A clearer indication of the effects of variations in the price of copper can be deduced from a previous simulation (see chapter 5) in which the LME price of copper increases by 10 percent. This considers a variation in price of the same magnitude and direction with respect to the actual historical price, for example, 10 percent above the actual price. An evaluation of the positive or negative impacts of such an increase or decrease in export prices could be used in analyzing the potential outcomes of price variability.

Most of the variation in domestic production is accounted for by the variability of value added in the mining sector, which in turn affects other sectors linked to copper production. However, cutbacks in gross noncopper investment induced by reduced government outlays bring down construction activity, thus offsetting the stimulative effect of an increase in value added by the Gran Mineria for some of the years.

The demand-sector variables show a fluctuating path with respect to the control-solution values. The aggregate demand variables have almost identical means with some variation in the corresponding standard deviations. The sample standard deviation for private-consumption expenditures is slightly lower with the smooth price of copper. For a number of years, real per capita net of taxes wage and nonwage incomes move in opposite directions, inducing greater cyclical behavior in private-consumption expenditures. The 2 years marked by historically high copper prices, 1966 and 1974, show significant increases in the rate of capacity utilization and depletion of inventories as demand adjusts to the sudden shifts in production induced by the copper sector.

The two principal linkages between copper price variability and domestic activity are through prices and the public-sector budget. The real current-account surplus/deficit of the government responds to fluctuations in the price of copper as well as price-induced variations in value added by the copper sector. In 1966, when the actual LME price was very high compared with the smooth price, there was a reduction in public savings necessitating additional financing. Yet the deterioration in the balance of payments draws down international reserves sufficiently to offset the expansion in Central Bank credit to the public sector and results in a lower value for the GDP price deflator. The most extensive fluctuations were recorded for the balance-of-payments surplus or deficit and the public-sector budget, bringing about greater price instability with a variable than with a smooth price of copper.

The results of the variable- versus smooth-price simulation show the benefits derived from a smooth path in copper prices. In chapter 3, an unsuccessful attempt was made to link domestic activity (for example, imports, government expenditures) to some measure of uncertainty originating in the export sector, as given by the standard deviation of total or copper export earnings. If the coefficients had been significant and exhibited the correct sign, there would have been some basis for the accepted view that perceived risk dampens or stimulates economic activity whenever the path of copper prices is variable or smooth. Thus the simulation in this chapter is based on a model in which these' risk factors do not operate explicitly.

Price Variability and Investment: An Alternative Simulation

As a test of the importance of export variability on overall risk, another simulation was performed with an alternative version of the noncopper gross fixed-investment equation. The specification is identical to the version discussed in chapter 3, except for the inclusion of the standard deviation of the GDP price deflator.

Because of the political changes and economic aberrations, such as the chronic rates of inflation, the risk factor may have evolved as a relevant consideration by decisionmakers, so that in addition to considerations of user cost of capital, rate of change of output, and capacity utilization, the price-variability measure was included in the investment equation and its coefficient was significant and negative. The impact of fluctuations in the price of copper would thus be felt directly through this variable. However, the standard deviation of the GDP deflator incorporates other sources of instability, such as political changes or economic policy decisions, that do not respond to developments in the copper sector. To the extent that these other factors can offset or intensify the fluctuations in export earnings, the effects of variability on investment could not be totally ascribed to export instability.

This alternative simulation was performed for the period 1965–1973. With the inclusion of the uncertainty term in the investment equation, the results show little if any differences in the average values between the variable and the

Table 6-4
Smooth versus Variable Price Simulation with Variability Term in the Investment Equation, 1965–1973
(*mean and standard deviation*)

	Variable Price[a]	Smooth Price
Gross domestic product (1965 pesos)	22.323 (2.071)	22.293 (2.066)
Private-consumption expenditures (1965 pesos)	16.616 (1.934)	16.542 (1.871)
Noncopper investment (1965 pesos)	2.985 (.184)	2.979 (.184)
Inflation	87.279 (153.212)	85.177 (146.851)

[a]The variable-price case is represented by a modified control simulation which incorporates the alternative investment equation.

smooth price of copper. Surprisingly, investment has the same value for the standard deviation in the smooth- as in the variable-price case. Whatever discrepancies occur between the other variables are due to the use of a modified control solution that incorporates the alternative investment equation. The results for several variables are displayed in table 6-4.

Conclusions

Although the smooth- versus variable-price simulation offers some interesting results, the analysis is by no means conclusive. If a more detailed national accounts were published that would be compatible with the pre-1968 disaggregated accounts, a more discriminating analysis of sectoral responses might reveal new insights.[12] The basic conclusion of this study is in agreement with MacBean's analysis that there are very small changes in real output. However, a variable price does bring out greater domestic price instability. This latter negative effect might be offset by the application of fiscal and monetary policies. The next chapter considers some of the stabilization mechanisms that might be available to policymakers.

With the copper sector, there are no definite advantages to a smooth price in terms of production levels. However, with the smooth price, the share of wage income out of total value added shows a significant improvement. For the noncopper sectors, one of the principal effects of export price variability is greater inflation. The latter is induced, in part, through the impact of international reserves and the financing of the public sector by the Central Bank, both of which enter directly in the determination of the monetary base.

Notes

1. Copper's share out of total exports of Chile is about 70 percent. See table 2-5.

2. For a current review of the literature, see Adams and Behrman (1981).

3. Bankers evaluate a country's credit rating to determine the interest rates and terms of the loan. A country's ability to pay may be questioned whenever its export earnings have fluctuated erratically.

4. MacBean, 1966, p. 68.

5. Rangarajan and Sundararajan (1976, p. 372).

6. Lira (1974) presents a good analysis of the linkages between the copper and noncopper sectors, but he does not deal directly with the issue of instability.

7. This is applicable to the pre-Chilenization and nationalization period (1926-1959), which is the period of MacBean's study.

8. MacBean, 1966, p. 184.

9. This variability index is defined as the sample standard deviation of X divided by its mean.

10. The purpose of the simulation is to study the impact of variability while holding constant the average level of prices.

11. In 1974, the LME price attained the highest average annual value during the period of simulation at 93.39 US¢/lb.

12. This would involve the kind of detailed econometric model that was estimated by Behrman (1977), but with emphasis on separating the copper and noncopper variables.

7

Copper Development and Stabilization Policies

The purpose of this chapter is to review some of the more interesting aspects of the simulations that have been presented in this study and to consider what policy implications can be derived from the analysis of their results. For example, in the last chapter, the simulation dealt with the consequences of price variability; primarily these were greater inflation and variability of domestic activity. The following sections present a discussion of alternative policy measures that could offset some of the negative impacts of export instability.

Whereas the copper-sector simulations have provided useful insights into the interconnections between that sector and the rest of the economy, this chapter deals with policies that could affect economic growth and stability. Some of the proposals have been derived from an analysis of particular simulations, for example, the multiplier effect of an increase in copper production (see chapter 5). However, these recommendations, or summary considerations, are not meant to be definitive; instead, they are general considerations of what types of orientations the authorities could pursue in expanding the role of the copper industry. The advantage in having used an econometric model is that the effects of various policies can be traced within a consistent economic structure, in contrast with the partial equilibrium analysis, which considers one or two variables and risks ignoring other relevant information.

The following sections will focus on the nationalized copper industry, which accounts for about 80 percent of total output. The leading questions will be (1) What should be the role of the government within the nationalized industry? and (2) What are the appropriate short-term stabilization policies that could neutralize the adverse consequences of the erratic behavior of copper prices? The task of evaluating policy alternatives is a relatively difficult one because of numerous caveats concerning the underlying assumptions. The latter are based on "reasonable" assumptions about exogenous variables in response to changes in the basic macroeconomic variables and would not necessarily materialize in actual fact.

The material presented in the previous three chapters is a collection of policy simulations that provide some quantitative information on the potential effects of various policy postures. This chapter brings together the basic elements of these simulations into alternative policy measures as supported by the individual numerical results (for example, growth rates, multipliers).

The long-run copper development strategies presented in the first part of this chapter encompass four important areas of decision making. The first considers the targets or objectives with respect to an expansion in copper output. The next policy consideration is the design of a copper-investment strategy that would lead to a realization of output targets and also stimulate economic activity in other sectors. The third area of policy decision concerns the management of copper revenues. Since the government owns most of the copper industry, the earnings from this highly profitable activity provide essential revenues that can be channeled into development projects. The fourth aspect of national development policies deals with the distribution of income. Economic growth means more real income; however, if this increase is not evenly distributed, some of the structural changes that are necessary to induce future growth may not materialize.

The second part of this chapter is concerned with the formulation of policies to cope with the inherent instabilities of the world copper market and their unfavorable impact on the Chilean economy. The implementation of some price-stabilization strategies serves to protect the domestic economy from the undesirable consequences of copper price variability, which is primarily realized through domestic price variability. These policies can be divided into two categories: medium- and long-term strategies, and short-term strategies. The first touches upon the long-term options in wrestling with the impact of copper price variability on domestic instability. These policies involve structural and institutional changes in the government, as well as in the copper sector. The second characterizes the standard fiscal-monetary policy instruments applied in the short run to offset changes in the equilibrium conditions in the goods and money markets brought about through fluctuations in export earnings.

National Copper Development Policies

This section outlines some of the important issues in the formulation of a comprehensive copper development plan. Without a deployment of too many details, the analysis will summarize the principal recommendations in the areas of production, investment, management of copper revenues, and income distribution. The success in reaping the maximum benefits from the abundant reserves of copper will determine the extent of the contributions to overall economic growth and development. Benefits cannot be measured solely by export earnings; the gains from copper exports will impact all other variables in the economy. Thus any decision concerning development of copper resources should be accompanied by an integrated plan to distribute the financial gains to the other sectors of the economy.

Expanding Copper Production

Chile has ample reserves of copper and is a low-cost producer; it should continue to exploit this resource. The theory of comparative advantage basically states that a country should specialize in the export of those goods which use intensively its most abundant factors of production. The extent of comparative advantage would be reflected in the relative costs of production. Chile's Gran Mineria is one of the world's lowest-cost producers of copper. In part, the cost advantage is derived from the high grade of its ore. The government, as owner of this resource, determines at what rate the ore should be extracted. This decision will eventually have an impact on the grade of the ore, since intensification of the mining process would lead to lower grades as the higher-quality grades are exhausted.

At present, the Gran Mineria contributes about 5 percent of GDP.[1] Because of the slow response of copper production to relative prices, most of the variations in copper prices would be felt through government-sector revenues and the overall price level. These represent the principal channels for transmitting external instability to the domestic economy.

If the public sector could implement some price-stabilization policies that would offset variations in copper prices, it would be conceivable that the share of copper in GDP could increase without inducing significant increases in domestic instability from fluctuations in export earnings. One drawback of this expansion could be a diminished employment effect, resulting from the large capital requirements relative to the rest of the economy. The capital-labor ratio for the Gran Mineria is about five and a half times as large as for the noncopper sectors (see table 2-8).

The differences between employment creation through an expansion in copper output and an increase in government expenditures can be established by comparing the former with the results of the simulation of a 10 percent increase in government-consumption expenditures. In the first year of the simulation, the increase in total employment (mining and nonmining sectors) is 11,000 for a 10 percent increase in government expenditures and 10,600 for a 10 percent increase in copper output.

Despite this employment disadvantage, the revenues generated through increased copper production could be channeled by the government into various labor-intensive development projects. Based on the decision of how much copper is to be exported, the authorities would determine the inflow of foreign-exchange revenues. The degree of success in channeling these resources into a viable industrial and agricultural development plan would determine the extent of the benefits from an expansion in copper exports. If the authorities decide to expand the level of copper exports, a part of these earnings could be invested in the priority sectors. It may be more desirable to use copper exports to finance imports of technology, capital goods, and raw materials to strengthen the domestic economy.

By increasing its production of copper, Chile could gradually raise its share of the world market. In 1977, its share of total world production was 13.2 percent. During the previous 5 years (1973-1977), this share had been increasing: 9.8, 11.8, 11.3, 12.8, and 13.2 percent, respectively. All Chile's output is now sold by CODELCO, which functions in some respects as an oligopolist in the world market. The five largest copper producers, including Chile, account for about 56 percent of total world output.

Whereas Chilean copper had been priced previously for the U.S. producers market, since the nationalization, price determination has been governed by CODELCO, which has sought to diversify the destination of its sales throughout the world. These are important considerations in evaluating the potential gains from producing more copper. These gains would depend to some extent on CODELCO's marketing strategies in the world market and its success in selling at the "right" price.

The potential benefits from an expansion in copper production in order to attain an influential position in the world market are very tenuous. One of the basic outcomes of the simplest model of oligopoly behavior is that anything can happen whenever one oligopolist does not anticipate how the others will react to a particular market situation. These uncertainties give rise in some cases to collusion or, as is more commonly known, cartels. If CODELCO wishes to expand copper production in substantial amounts, these actions might be coordinated with the other major producers; otherwise the increased supply of copper would depress the world price.

However, if Chile had raised production of copper by 10 percent above its historical value every year between 1968 and 1976, the impact on the world price would be minimal. This simulation was elaborated using a two-way (or simultaneous) model of the world copper market and the Chilean economy.[2] The average yearly decrease in the LME price during that period as a result of the Chilean increase in exports was 1.3 percent. The simulation of a 10 percent increase in production without a price effect (as described in chapter 5) generated an increase in copper export revenues on an average of 10.39 percent per year with respect to the control solution between 1968 and 1976; allowing for the world price effect through the linked models, the corresponding figure is 9.32 percent.[3] The difference is almost all attributed to the minimal change in price.

The development of copper-production policies should be guided by price considerations as well as the multiplier effect, the impact on government revenues, and the desirability of channeling investment funds into the copper sector. The simulation of a 10 percent increase in copper production revealed an impact multiplier of 2.8 with respect to GDP. The dynamic effect reflects the degree of interdependence between copper and the rest of the economy. The indirect linkages to production and employment occur through the provision of domestic inputs for the Gran Mineria. The amount of foreign exchange above the control-simulation value generated by the Gran Mineria during the 12-year simulation is US$892.51 million. The 10 percent increase in copper production also created more demand for labor; in the Gran Mineria, employment increased on average 8.7 percent per year, and in the noncopper sectors, by 1.9 percent.[4] In addition to the favorable income-multiplier effects, further increases in output would add to current-account revenues of the public sector. The surplus on the current account of the public sector measured in constant 1965 pesos is improved by about 6.5 percent per year. The increase in revenues could contribute to the financing of essential government investment in infrastructure.

The principal conclusion pertaining to an expansion in copper production above current levels is that it is advantageous. However, it is not clear whether pursuing an ambitious world-market strategy would yield a favorable outcome. The relative insensitivity of the world-market price of copper to a moderate increase in production was demonstrated by the two-way simulation, using the Chilean econometric model, linked simultaneously with the world copper model, so that the market can tolerate a small increase in Chile's share without straining the earnings of the other major exporters. An analysis of the indirect employment and income effects also indicates that growth in the copper sector will affect overall growth and development.

Benefits from Copper Investment

Any plans to expand production must incorporate a feasible investment program unless the mines are operating at very low levels of capacity. In deciding on the size of investment expenditures necessary to sustain output targets, the government must consider four issues: the investment multiplier, the generation of government revenues, the redistribution effect, and the type of technology. More investment in the Gran Mineria can stimulate domestic activity. The impact would materialize through employment and purchases of machinery and equipment, especially when the project involves major construction activity. The immediate impact on aggregate demand will occur through construction activity and imports of goods and services (machinery and equipment). Within the production sector, manufacturing, public utilities, and services will respond to the increase in aggregate demand. The increase in domestic activity will generate more employment. As the copper investment simulation brought out, the dynamic multiplier effects are very similar to the case of noncopper government investment. However, the copper-investment program will have a more favorable impact on the balance of payments, and as a consequence, it will improve the country's international credit rating. In the case of an increase in government-investment expenditures, the average change in the balance of payments was -19.5 percent, while for an increase in investment in the copper sector, the change was +16.1 percent. With copper investment, the impact on overall domestic prices is more favorable because of the reduced financial burden of the public sector.

The copper industry represents one of the public sector's more profitable businesses (of course, during the low price years this may not be true). Perhaps an analogy could be drawn to Mexico's government-owned oil industry and its crucial role in the growth and development of that economy. The expansion in Gran Mineria investment will improve the government finances. As the increases in production are materialized, the public sector will augment its income from profits and indirect taxes in the copper industry. These revenues could be channeled into the less lucrative and in many cases deficit-ridden activities of the government whose social value may justify their continued operations. A high-priority area might be the strengthening of the infrastructure to sustain private-sector growth in agriculture and industry.

The income-redistribution and technology issues of copper investment are concerned with the socially acceptable allocation of income between the factors of production. The investment in copper will bring about a redistribution of income in favor of the nonwage income component within that sector. A larger share of nonwage income would be reflected in more government earnings. The public sector might ameliorate the situation by granting a more liberal wage policy. Since the wages in that sector appear to average above wages in the rest of the economy, the increased earnings from copper investment could be distributed to the workers in the form of government bonds (worker participation in profits) or through the provision of amenities in the living conditions of the communities surrounding the mines (for example, special training facilities, medical services).

A more desirable approach to income allocation is through the selection of various techniques that favor the substitution of domestic for foreign technology. CODELCO has achieved much success in this area by requiring that individual mines submit their contracts with foreign suppliers for approval. If there are no domestic products of comparable quality, the mines are allowed to import the materials. This policy has contributed to a gradual change in the structure of material inputs. In the period 1960-1962, domestic material inputs represented 54.1 percent, and in the period 1974-1976, 71.4 percent.[5]

Thus far, the purchase of capital goods has not been subjected to regulation by CODELCO. The extension of current policies to encourage the purchase of domestically produced machinery and equipment would improve Chile's share in the provision of factor inputs. The backward linkages thus created would stimulate employment in other sectors, and the multipliers yielded by the simulations would be somewhat bigger.

Management of Copper Revenues

Before nationalization, the copper tax legislation had represented the principal channel for recuperating the industry's profits. Now that the government owns this valuable resource, its tax policies can promote more efficient methods of production. The traditional tributary instruments have been production, over-price, and profits taxes. The tax rates could be maintained sufficiently high without discouraging investment in the copper sector. Regardless of the needs for foreign exchange, the use of an overvalued Gran Mineria exchange rate should be avoided. The results of the simulation showed that an overvalued Gran Mineria exchange rate led to higher costs of production and a reduction in copper output.

Recently, some large multinational companies have taken advantage of the liberal foreign-investment code in Chile.[6] The attraction of foreign capital has been necessary, particularly because of the large investment requirements for the development of new deposits. After the rough experiences of Chilenization and nationalization, the government might proceed with a firmer posture in extracting the maximum surplus from the foreign investors.

In view of the importance of copper revenues and the recent change in ownership of the copper mines, the public sector may benefit through a reorganization of its budget appropriations. Copper revenues could be treated differently. The estimation of expected revenues could be calculated with a "normal" price of copper.[7] This index would not be subject to the short-run aberrations of the market that can generate sudden wealth or ruinous losses. Whenever the price is significantly above the normal level, the excess over anticipated revenues could be set aside through some form of financial investment or applied toward the reduction of the external debt. The impediments to the implementation of this policy could arise from short-term political pressures. Moreover, by use of a normal price of copper, the public sector may have to depend on adjustments in noncopper tax rates to offset increases in total outlays whenever the actual price is below the "normal" level.

Tironi (1977) suggests the establishment of a copper-stabilization fund. The administration of this fund is described as follows: "The manager of the Copper Fund should have explicit authority to invest in short-term securities at home and abroad, and to make loans or buy bonds of the public and private sector."[8] The purpose of this fund is to serve as a buffer against sudden changes in the price of copper. By using the proceeds from the high-price years to offset the shortfalls in the low-price years, the administrators of the fund could stabilize the flow of foreign exchange and, hence, the domestic price level.

A simulation of the model showing how the fund might operate during the 1965-1976 period is very difficult to design. This type of analysis is best done for 1 year (the second part of this chapter considers a 1-year offsetting policy). Because of the dynamic properties of the model and, hence, the economy, the results of the simulation are dependent on whether the price in the first year is above or below the smooth price. If the price during the first year is below the smooth price, then the application of fiscal and monetary policies for the following years would be expansionary, in order to stimulate aggregate demand. These expansionary policies would be implemented throughout the period of simulation even though the actual price is above the smooth price in some of the subsequent years.

Income Distribution

Income redistribution has been a policy priority during past administrations; however, the most concerted effort was realized during the Allende period. Because of the importance of income distribution in determining social and economic stability, the government might incorporate in its long-term policies some plans for achieving a more even distribution of income. A more equitable distribution cannot be defined ambiguously; a highly uneven distribution can give rise to a political catastrophe with grave consequences for the economy. Whenever the economy prospers from a copper bonanza, the government could earmark some of the increased earnings for those areas which may positively affect a structural shift in the allocation of the national wealth.

The material from different simulations in chapter 5 reveals some interesting information about the potential changes in the income distribution.[9] With an increase in copper production, there is a minimal increase in the average share of wage out of total income in the Gran Mineria. For the control solution, the average share is 37.1 percent, and for the simulation of a 10 percent increase in production, it is 37.6 percent. The smooth copper-price simulation showed the most significant change, with an increase to 43.3 percent in the share of wage out of total income. The noticeable improvement is explained by greater price stability in the copper sector. The positive effect on income distribution from an expansion in copper production supports a policy stance favoring a larger share of copper in GDP, and the benefits derived from the smooth-price simulation bear out the importance of adopting some of the long-term stabilization strategies to be discussed in the following section.

While an increase in the LME copper price induces an expansion in domestic activity, there is a deterioration in the distribution of income as the average wage share in the copper sector drops to 34.7 percent from the control solution's 37.1 percent. An expansion in copper investment has a similar unfavorable effect; in this case, the average wage share declines to 36.2 percent. Both the price and the investment increases favor the nonwage income component. Since the government is the principal owner, it could use a part of the increase in revenues to improve the distribution of income within that sector.

Copper-Market Fluctuations and Policy Responses

The purpose of this section is to discuss some of the policies that might be employed to stabilize the domestic economy in response to fluctuations in the world copper market. The growth targets for the copper industry and the investment decisions needed to achieve those levels of production should be made relative to the objective of maximization of economic benefits to all sectors. This decision process may be one phase of the overall development planning. The other, which is important for the success of the first, is what policies should be implemented to offset any unexpected changes in world-market conditions that might jeopardize the success of a copper development plan. In view of the volatile conditions in the world copper market, what defenses can the government rely on to ensure the success of its plans?

Medium- and Long-Term Strategies

From the analysis presented in the previous chapter, it is not obvious that international price-stabilization schemes are necessary. It is not unlikely that individual countries could fend for themselves. In the case of Chile, the government may be able to offset some of the potentially disruptive outcomes in the balance of payments. After identifying the secular trends in the terms of trade, the government might be in a position to consider alternative schemes to offset the unfavorable trends. One approach could be a more rigorous management of long-term external debt. The balance-of-payments constraint affecting LDCs usually arises from too much dependence on commodity exports with unstable prices. By concentrating long-term planning in the area of external debt management, some of the uncertainty associated with foreign-exchange earnings might be alleviated. The objective is to neutralize the impact of long-run fluctuations in the world market. Some of the available policy instruments to accomplish this goal include accumulation or sterilization of international reserves, control over the world market, accessibility to external credit, and coverage of financial risk.

This section will briefly consider some of the costs and benefits that might be associated with each of the four strategies. The selection of a particular policy option would depend on the position of the economy within the normal cycle (for example, a recession) in addition to the world copper-market conditions. While each alternative would involve the application of different policy instruments, the complexity of the national and international economic environment would call for a combination of some or all of the four strategies.

Perhaps the simplest alternative is the accumulation and sterilization of international reserves. During the good years, the Central Bank would accumulate reserves accompanied by measures to reduce the expansionary effect on the monetary base. This option has been discussed in the previous section. However, this approach may falter in view of the interrelation between the Central Bank and the public sector, especially as it applies to the financing of the budget deficits. If the government avoids using unexpected windfall earnings of foreign exchange, these resources might be applied toward a reduction of its external debt.

With the government as the owner of the industry, it may be in an influential position in the world market. However, the potential effect on the world price of copper is uncertain, since this option has not been exercised previously. If the members of CIPEC can come to a unanimous agreement over official policy, their control over the world's supply may lead to a favorable outcome. The success of a copper cartel hinges on two considerations: one is how long could the organization adhere to a high-price policy without bringing about the disintegration of its membership; and second, what would be the likelihood that other metals would be substituted for copper in the long run, if the real price rises above some threshold? As the simulation in the previous section brought out (by linking the world copper-market model with the Chilean model), an increase of 10 percent in copper exports would have a small impact on the world price. Based on these results, it may be questionable whether Chile alone could influence the world price. Thus a group of the principal exporters might be more effective.

Another option is to rely on external credit to make up the loss in foreign-exchange earnings (the following section considers the case of a 1-year shortfall in export earnings). The public sector is responsible for a significant proportion of total gross domestic investment and has been able to acquire external financing for much of the imports of machinery and equipment. By pursuing this policy, the government could encounter the undesirable situation of a deficit in the balance of payments that would necessitate the use of the limited, as well as more expensive, short-term credits. The availability of external financing is subject to credit limits imposed by international banks. A policy of early debt repayments could enable the Chilean government to build up a credit buffer to guarantee a stable inflow of essential material imports during the low-price years. With the exception of some temporary disruptions in imports resulting from shortage of foreign exchange, the reliance on external financing provides a means for the economy to spread its risks over time.

Finally, in addition to the standard credit channels as a means of compensating for future risks, there is the potential for other measures as a hedge against the unexpected variations in the price of copper. Lessard (1977) provides an interesting analysis of how, with relatively free international capital markets, a country can trade risky for less risky assets. One example is the use of commodity-denominated bonds. Chile could issue certificates denominated in metric tons of copper and bearing a fixed rate of interest. The value of the bond would be calculated by multiplying the current price of copper by the volume backing up the note. If, at the time of redemption, the value is greater than its purchase price less accrued interest, the individual would receive the difference. These notes would serve as a mechanism for transferring some of the risk to the international financial markets. However, the implicit interest cost may be higher than in the case of external bank-credit channels.

The Chilean government's current exchange-rate policy represents another means of hedging against future uncertainty. By announcing the official exchange rate 12 months in advance, the Central Bank provides some indication about future expectations and thus assists domestic firms in planning their operations for the upcoming year.

Short-Term Strategies

While the medium- and long-term strategies are mainly concerned with the establishment of mechanisms that act as buffers against erratic fluctuations in copper prices, in the short run, cyclical behavior necessitates in some instances the use of stabilization policies. Through the implementation of fiscal and monetary policies, the government could guide the economy toward a situation of "equilibrium" in the goods and money markets. The principal impediments to the attainment of stability are the existence of market imperfections, changes in external events, and government-imposed development objectives that require continual intervention in the market. Because events in the world copper market are uncontrollable and unpredictable from the point of view of Chile, short-run fluctuations in the price of copper could prevent the achievement of long-run stability if the government does not have the flexibility to respond with appropriate monetary and fiscal policies.

The instruments of fiscal policy include control over government expenditures, tax levels, and structure; transfers and subsidies; and balance-of-payments restrictions, such as import tariffs and quotas, as well as access to external credit. The choice of instruments depends on the structure of the political institutions, as well as the economic situation. Government expenditures can respond more quickly to stimulate aggregate demand than changes in the tax rates, which would involve a lengthy process of legislative approval. Since public-sector expenditures include the payment of wages, the hiring of employees, and the purchase of domestic and imported goods in almost all sectors of the economy, this variable represents an important instrument of stabilization policy. For instance, the distribution of income can be affected through higher government wages or an expansion in employment. Government wages also set a standard for wage negotiations in the private sector.

Monetary-policy instruments comprise Central Bank credit to the government, the reserve-requirement regulation, and changes in international reserves. The use of Central Bank credit to the government is constrained by the dependence of the latter, in view of the somewhat rigid budgetary allocations, on financing from the monetary authorities. In Chile, the Central Bank has acted in a somewhat passive role as the succor of the public purse.[10] In contrast, the reserve-requirements policy is a relatively flexible mechanism for controlling the capacity of commercial banks to expand the money supply. Finally, changes in the level of international reserves affect the ability to import and, in conjunction with a consideration of the internal-external inflationary gap, are used by the authorities as indicators in determining the official exchange rate. One of the crucial balance-of-payments variables that is directly controlled by the authorities is the flow of medium- and long-term credits. Fiscal policy affects the balance on current account through import tariffs or quantitative restrictions, whereas monetary policy affects the level of foreign-exchange reserves through the access to external credit.

The application of the two types of policy instruments should be coordinated in response to the short-term stabilization objectives. Because of the importance of copper exports, the ability to manage the international monetary variables, such as the short- and long-term flows of foreign exchange, is an essential element in domestic price stabilization.

Model Simulations and Short-Term Strategies

The importance of short-run policy responses to unanticipated fluctuations in the international copper market can be appreciated through the use of model simulations depicting a 1-year increase or a 1-year decrease in the LME price of copper. These simulations would be similar to those in the price-variability chapter with the added advantage of isolating the impact through time of a price increase or decrease on the rest of the economy. In the case of a 1-year decrease in the price of copper, two simulations were performed. The first shows the impact on the economy without government intervention. The second assumes that the authorities would adopt some measures to offset the more harmful effects. For each case, the basic chain of events is described followed by a brief indication of some feasible policy responses.[11]

If the LME price of copper dropped by 10 percent in 1 year, the GDP price deflator would be about 6 percent below and real GDP would be about 1 percent below the control solution (see table 7-1). The immediate impact is a deterioration of international reserves by 45 percent and a drop in the real current-account savings of the public sector of 10 percent. In the first year, the decline in the GDP price deflator is explained by the drop in the level of international reserves that offsets the increase in the flow of credit to the public sec-

Table 7-1

A 10-Percent Decrease in the Price of Copper in 1 Year

(*percent deviation from control solution*)

	1965	1966	1967	1968	1969
Production (1965 pesos)					
Gross domestic product	-.06	-.99	-1.05	-.81	-.67
Gran Mineria	-3.61	-.66	-.73	-1.10	-.53
Demand (1965 pesos)					
Private consumption	.28	-.54	-1.11	-1.44	-.96
Public consumption	.44	-.35	-.71	-.45	-.52
Non-Gran Mineria investment	.38	-1.04	-1.38	-1.71	-.99
Imports of goods and services	.47	-1.95	-1.25	-.97	-.80
Public sector (1965 pesos)					
Current savings	-9.75	-9.01	.12	1.68	-.75
Flow of credit to public sector	22.88	25.66	-1.47	-6.98	3.66
Money and Prices					
Balance-of-payments surplus or					
deficit (mill. US$)	-44.90	7.46	11.06	2.96	3.11
Money supply	-6.92	7.31	9.34	4.81	6.05
GDP price deflator (1965 = 100)	-5.89	1.81	4.69	1.16	2.83

tor. During the following 3 years, there is a continued decline in real GDP induced by the original decrease in the price of copper. Because of a reduction in international reserves, imports of goods and services are 2 percent lower in the second year. Also, the 10 percent reduction in government savings leads to a 1 percent decline in non-Gran Mineria investment. The combination of Central Bank financing of the public-sector deficits and the increase in international reserves brought about through reduced imports gives rise to an increase in the GDP price deflator after the first year, which in turn affects private-consumption expenditures and GDP.

The proper policy response in the preceding situation might include an expansion of public-consumption expenditures. The multiplier effects of an increase in expenditures were detailed in chapter 4. This increase in government outlays could be financed through external credit and thus provide the needed foreign exchange. This will also allow the government to maintain investment expenditures at previously committed levels. There is also the possibility of granting some relief through a lower reserve-requirements ratio, permitting commercial banks to extend greater amounts of credit to domestic producers, or an increase in reserve requirements in order to finance the deficit of the public sector. This latter approach would place the burden of a reduction in the price of copper on the private sector, while the stimulative spending of the government would strengthen aggregate demand.

In order to derive the possible range of magnitudes that may be required to offset the impact of a 10 percent reduction in the price of copper, an alternative simulation was considered. The basic assumption was that the authorities would rely on consumption expenditures and foreign financing to offset the decrease in GDP from the decline in the price of copper. As shown in table 7-2, the increase in government expenditures is 9 percent, or .176 million × 1965 pesos. This increase in expenditures is financed through external borrowing of US$58 million during the first year. It is assumed that all additional financial requirements of the public sector are met through foreign borrowing, so that the flow of credit to the public sector remains unchanged with respect to the control solution. The results of this simulation indicate the feasibility of adopting fiscal and monetary policies to stabilize the domestic economy whenever copper prices decrease or increase.

In the event of a 10 percent increase in the price of copper, the only unfavorable impact occurs through the GDP price deflator. During the first year there is an increase in international reserves of 43 percent. Nonwage income in the copper sector increases by 25 percent, capturing most of the increase in revenues. The increase in nonwage income raises government tax revenues, and hence, real current-account savings of the public sector increase by 9 percent. The expansion in international reserves is greater than the reduction in Central Bank credit to the public sector, so the change in the money supply brings about an increase in the GDP price deflator of 4.4 percent. During the second year, imports of goods and services increase as a result of greater availability of international reserves, and the increase in government savings induces more investment. The

Table 7-2

A 10-Percent Decrease in the Price of Copper in 1 Year Accompanied by Offsetting Policies

(*percent deviation from control solution*)

	1965	*1966*	*1967*	*1968*	*1969*
Production (1965 pesos)					
Gross domestic product	.04	-.36	.23	.01	.40
Gran Mineria	-3.61	-.04	-.24	-.03	-.80
Demand (1965 pesos)					
Private consumption	-.43	.02	-.45	-.63	.71
Public consumption	9.00	15.10	8.91	5.38	3.51
Non-Gran Mineria investment	.62	.04	-.49	-1.13	-.60
Imports of goods and services	.36	-.19	.39	-.19	-.33
Public sector (1965 pesos)					
Current savings	-6.48	-16.19	-4.76	-7.21	-2.68
Flow of credit to public sector	0.0	0.0	0.0	0.0	0.0
Money and Prices					
Balance-of-payments surplus or deficit (mill. US$)	-3.22	-3.78	9.36	-1.03	-3.08
Money supply	-1.67	-3.29	-5.67	-6.54	-7.93
GDP price deflator (1965 = 100)	-.76	-1.96	-3.91	-4.70	-4.77

most significant effect on real GDP occurs in the second year, with a 1.7 percent increase.

While the copper-price increase led to a rise in real income and domestic activity, this had some less desirable consequences, as seen by the increase in the general price level. One of the options for stabilizing domestic prices was discussed in the previous section on medium- and long-term strategies. This is a sterilization of the excess foreign reserves occuring during the first year. These funds could be used to retire part of the external debt, and the remaining portion could be placed in the international capital markets as a reserve for those years experiencing an unanticipated drop in the world price of copper. However, this latter option could be severely constrained if the government faces indomitable pressures to invest the export earnings on the high-priority sectors, such as agriculture, housing, and education, instead of investing abroad.

Conclusions

The material presented in this chapter is not intended to exhaust all possibilities, but rather to provide a brief overview of some general policy areas. Some of the proposals are based on the results obtained from the simulations in chapters 4, 5, and 6. The economic implications of some of the policy suggestions can be appreciated through the quantification of the relevant variables in each of those simulations.

The results of the price-variability analysis are instructive, but more research is warranted. Despite the extensive disaggregation of the model, the availability of adequate data by sector could permit a more detailed estimation of the supply side and thus enhance the senisitivity of the simulations. In general, the characteristics of the model have been amply tested, showing only minor peculiarities, such as the exogeneity of long-term capital flows. Even if some of these variables were respecified, the basic conclusions would not be altered. The policy options discussed in this chapter are valid measures to counteract the undesirable consequences of variations in copper earnings that have been revealed through the model simulations.

As demonstrated in the first part of this chapter, the government, as the owner of that industry, may benefit from an expansion of production. The type of investment policies adopted will determine how much domestic employment can be generated. Once the desirable level of copper output is determined, the stabilization measures in the short, medium, and long run could involve the accumulation and sterilization of international reserves or the issuance of commodity-denominated bonds. Because of the relatively small size of the copper sector with respect to GDP, balance-of-payments and debt-management measures present the more appealing course of action to minimize the disturbing effects of export price instability.

Table 7-3
A 10-Percent Increase in the Price of Copper
(*percent deviation from control solution*)

	1965	*1966*	*1967*	*1968*
Production (1965 pesos)				
Gross domestic product	.19	1.72	.82	.22
Gran Mineria	3.21	.85	−.30	−.12
Demand (1965 pesos)				
Private consumption	.51	2.14	1.08	.16
Public consumption	−.07	.93	.92	.49
Non-Gran Mineria investment	−.48	1.43	.57	.66
Imports of goods and services	.12	2.96	1.32	.41
Public sector (1965 pesos)				
Current savings	8.60	8.01	−4.01	−4.18
Flow of credit to public sector	−19.99	−22.53	14.79	16.60
Money and Prices				
Balance-of-payments surplus or deficit (mill. US$)	42.73	−12.53	12.22	−1.65
Money supply	6.80	−8.62	−6.04	.37
GDP price deflator (1965 = 100)	4.37	−3.52	−4.70	−.30

Notes

1. See table 2-1.

2. The world copper-market model was solved simultaneously with the Chilean model. See Pabukadee (1979) for a description of the copper model.

3. There are some differences between the control solution of the Chilean econometric model and of the joint system (Chile and copper).

4. During the period 1965-1976, employment by the Gran Mineria, represented, on average, 1 percent of total employment.

5. See table 2-7.

6. The most significant project to date is an investment of US$100 million by Exxon to purchase one of the medium-scale mines, the Mineria Disputada.

7. In the area of development planning, Tironi (1977) suggests that the government use some notion of a "normal" price that can be defined as a 5-year moving average.

8. Tironi, 1977, p. 38.

9. The discussion deals with the factoral distribution, since there are no published data on familial or any other type of income distribution.

10. See Wachter (1976).

11. Because the emphasis is on policy implications, the detailed description of the simulation has been left out. Some of the important results are similar to those described in chapters 5 and 6.

8 Conclusion

In the introduction to this study, two goals were defined: the specification of a detailed econometric model incorporating direct as well as indirect linkages to the copper sector, and the application of the model to analyze various copper scenarios to serve as a basis for drawing up copper policy recommendations. Both objectives have been elaborated through the material presented in the previous six chapters. The copper simulations in chapters 5 and 6 provide quantitative information that contributes to an understanding of some of the magnitudes of the effects on the Chilean economy of varying the assumptions about copper-sector variables. The analysis of chapter 5, dealing with copper-sector variables, has been useful in drawing up some of the basic elements of a copper development plan in chapter 7. A principal conclusion derived from the numerical results of the simulations is that the Chilean government should pursue a copper-expansion strategy. The increased revenues could be allocated to the development of other less profitable, but socially desirable activities.

The export-instability analysis has dealt with price variability. Since the world-market price of copper reflects the precarious relationship between world suppliers and consumers, its fluctuations are beyond the control of Chilean exporters. The attainment of desirable targets of economic growth, distribution, and stability will be contingent on the skill of policymakers in offsetting the disruptive effects of export-price instability. While the impact from copper-price variability is realized through greater domestic price instability, the implementation of prudent monetary and fiscal policies could prevent many of the undesirable consequences. However, more econometric research in this area is needed to confirm these conclusions.

One area for further development is the disaggregation of the model. The simulations showed that by providing greater detail in the linkages between the copper sector and the rest of the economy, more so than in the Lira (1974) model, a more significant impact from any changes in the copper-sector variables (for example, production, investment, and so forth) is demonstrated. Future work in this area might involve the specification of a model with at least ten production sectors and greater disaggregation of the demand-side variables. The feasability of such a model for Chile would in the future depend on the availability of additional detailed data.

One final caveat concerns the use of econometric models for the evaluation of alternative scenarios. The methodology pursued in this study has certain limitations. This is why the results of the simulations are not meant as numerically precise measures of multipliers or of other quantitative changes. For each simulation, one or more exogenous or endogenous variables have been changed while holding constant all other exogenous variables. The importance of those other variables in offsetting the particular assumed changes would determine the validity of the resulting numbers. In this study, the possibility of changes in other exogenous variables has been incorporated into the simulations; however, these added assumptions do not represent all the contingencies that could arise in the real world. In some cases, these unanticipated changes in some exogenous variable that is not accounted for within the simulation could dampen or intensify the initial exogenous "shock" applied to the system.

Appendix A:
Listing of Equations
of the Chilean
Econometric Model

This appendix contains a listing of the more important equations of the Chilean econometric model. The latter consists of 48 behavioral equations and 206 identities and exogenous variables. The model is divided into eight sectors:

1. Production sector
2. Domestic demand sector
3. Copper sector
4. Foreign trade sector
5. Price sector
6. Government sector
7. Monetary sector
8. Factors of production

The period of estimation varies with some of the equations; however, the majority of the time series are from 1960 through 1976. Variables expressed in real terms (with the suffix R) are denoted in 1965 pesos. The behavioral equations have been estimated using the ordinary least-squares method.

The principal source of information is the national accounts of Chile. Since there is no definitive statistical publication on the copper sector, various sources were combined in deriving a consistent set of data. All the information was obtained from official Chilean publications, except in a few cases in which secondary sources were used (for example, import price indices are computed from International Monetary Fund and OECD export price indices of main trading partners). The following is a list of the main sources of information for the copper and noncopper sectors:

Copper-Sector Data Sources

Banco Central de Chile. *Balanza de Pagos de Chile* (various issues).
Copper Studies, Inc. 1978. *Copper Studies.*
Corporacion del Cobre. 1974. "Tendencias de los Tipos de Cambio Periodo Enero 1970–Abril 1974."
Corporacion del Cobre. 1975. *CODELCO: Anuario.*
Corporacion del Cobre. 1977. *1ᵃ Memoria Anual 1976.*

Corporacion del Cobre. 1978. *2ª Memoria Anual 1977.*
Ffrench-Davis, R., and E. Tironi (Eds.). 1974. *El Cobre en el Desarrollo Nacional.*

Noncopper-Sector Data Sources

Banco Central de Chile. *Balanza de Pagos de Chile* (various issues).
Banco Central de Chile. *Boletin Mensual* (various issues).
De Castro, S. 1978. "Exposicion Sobre el estado de la Hacienda Publica: Presentada por el Ministro de Hacienda Sr. Sergio De Castro Spikula." Statistical appendix in *Boletin Mensual.* Banco Central de Chile.
Ffrench-Davis, R. 1973. *Politicas Economicas en Chile 1952-1970.*
International Monetary Fund. *International Financial Statistics* (various issues).
Oficiana de Planificacion. 1977. *Cuentas Nacionales de Chile 1960-1975.*

Production Sector

Real Value Added, Agricultural Sector

$$X1AGR = -1.4305 + 0.0057PX1AG/PX2MFG(-1) + 0.0315KNGMR(-1)$$
$$\quad\;\;(-1.89)\quad\;(1.74)\qquad\qquad\qquad\qquad\quad(3.45)$$

$$+ 0.6415X1AGR(-1) - .03573DUM72/73$$
$$(2.72)$$

$\bar{R}^2 = 0.778$
$D.W. = 2.39$
$SEE = .88814E-01$
Period of fit = 1961-1976

Real Value Added, Manufacturing Sector

$$LX2MFGR = -1.0629 + 0.7476LCER + 0.4729LIBR$$
$$\quad\;\;\;(-6.55)\quad\;(10.76)\qquad\;(5.72)$$

$$+ 0.0458DUM71/73 - 0.1196DUM75$$
$$(1.61)\qquad\qquad(-2.72)$$

$\bar{R}^2 = 0.970$
$D.W. = 1.17$
$SEE = .34945E-01$
Period of fit = 1960-1976

Real Value Added, Non-Gran Mineria Mining

X2MINNGMR = -1.6976 + 0.0431KNGMR(-1) + 0.4026X2MINNGMR(-1)
                (-2.20)   (2.32)   (1.57)

$\bar{R}^2 = 0.931$
D.W. = 1.59
SEE = .86881E-01
Period of fit = 1961-1976

Real Value Added, Construction

X2CONTR = 0.0716 + 0.3849IBNMKR
                (0.83)   (10.02)

$\bar{R}^2 = 0.889$
D.W. = 1.91
SEE = .45576E-01
RHO(1):0.544
Period of fit = 1960-1976

Real Value Added, Public Utilities (Electricity,
Gas Water, and So Forth)

X2UTR = -0.1418 + 0.0002CUPRODGM + 0.0141X12NGMUTR
            (-2.95)   (2.59)   (2.97)

                      + .06887X2UTR(-1)
                          (6.06)

$\bar{R}^2 = 0.973$
D.W. = 2.20
SEE = .17754E-01
Period of fit = 1961-1976

Real Value Added, Services Sector

X3R = 0.3806 + 0.3201C+I + 0.4613E+MGSR + 0.0012CUPRODGMAV2
        (1.00)   (12.74)   (4.46)   (1.29)

$\bar{R}^2 = 0.933$
D.W. = 1.84
SEE = .14232
Period of fit = 1960-1976

Domestic Production of Fuel, in Constant Pesos

PRODFUELR = 0.0178 + 0.0354X12R - 0.0122TREND
$\qquad$ (0.32)   (5.02)   (5.55)

$\bar{R}^2 = 0.645$
D.W. = 1.52
SEE = .22737E-01
Period of fit = 1960–1976

Real Value Added, Secondary Sector

X2R = X2CONTR + X2MFGR + X2UTR + X2MINNGMR + X2GMR

Current Value Added, Secondary Sector

X2N = X2CONTN + X2MFGN + X2UTN + X2MINNGMN + X2GMN

Domestic Demand Sector

Real Inventory Investment

ICHR = .7319 + 0.1906S - 0.05979DS - 0.4818INVR + 0.0006PGDP*(-1)
$\qquad$ (1.04)   (1.40)   (-0.56)   (-1.30)   (-1.00)

$\bar{R}^2 = 0.503$
D.W. = 1.53
SEE = .120886
Period of fit = 1961–1976

Real Gross Domestic Sales

S = GDPR - ICHR

Real Non-Gran Mineria Gross Investment

IBNGMR = 0.1652 + 0.0412DK*(-2) + 0.8920IBNGMR(-1)
$\qquad$ (0.33)   (2.01)   (2.27)

$\qquad$ + 0.0327DGDPNGMCAPUT + 0.0493GVSVR(-1)
$\qquad\qquad$ (0.73)   (0.73)

$\bar{R}^2 = 0.740$
$D.W. = 2.00$
$SEE = .17753$
Period of fit = 1963-1976

Desired Stock of Capital

$$K^* = 1.91613(PGDP/C)^{.36108} \times X12NGMR$$

Price of Capital Services

$$C = PIB \times [DEPRC + (RUSBLTN\$/100)]/(1-TXRATECORP)$$

Real Government Consumption Expenditures

$$CEGR = -1.1434 + 0.1017GVRVR1 + 0.1863NP1 + 0.5549CEGR(-1)$$
$$(-1.47) \quad (2.69) \qquad\qquad (1.27) \qquad\quad (2.84)$$

$$+ 0.1959DUM71/73$$
$$(3.17)$$

$\bar{R}^2 = 0.981$
$D.W. = 2.10$
$SEE = .77822E-01$
Period of fit = 1961-1976

Real Private Per Capita Consumption Expenditures

$$CEPR/N = 0.1442 + 0.7340WGSTOTNTR/N + .5603\ OFPNNT/IN$$
$$(2.46) \quad (9.56) \qquad\qquad\qquad (5.05)$$

$$+ 0.3180CEPR/N(-1) + 0.0001PGDP^*(-1)$$
$$(3.94) \qquad\qquad (2.49)$$

$\bar{R}^2 = 0.984$
$D.W. = 1.91$
$SEE = .21116E-01$
Period of fit = 1961-1976

Real Depreciation

CCAR = 0.6131 + 0.0234KR(-1) + 0.2115DUM71/76
         (1.03)   (2.03)   (1.71)

$\bar{R}^2$ = 0.727
D.W. = 1.95
SEE = .13139
Period of fit = 1960–1976

Real Non-Gran Mineria Capacity Output

GDPNGMCAP = .257777(.55 × KNGMR$^{-1.7695}$

$$+ .45 \text{ NPENGMMAX}^{-1.7695})^{-.56513}$$

Real Capital Stock

KR = KR(-1) + IBR – CCAR

Current Gross Domestic Product

GDPN = X1AGN + X2N + X3N

Real Gross Domestic Product

GDPR = X1AGR + X2R + X3R

Copper Sector

Realized Export Price of Copper, Gran Mineria in Dollars

PECUGM$ = 13.6228 – 10.3447DUM60/67 + 0.6576PCULME$
              (2.12)   (-3.19)   (6.91)

$$+ 0.0237\text{PCULME\$}(-1)$$
$$(0.26)$$

$\bar{R}^2$ = 0.913
D.W. = 1.78
SEE = 4.9571
Period of fit = 1960–1976

*Realized Export Price of Copper, Medium- and Small-Scale
Mining in Dollars*

PECUPMM\$ = 3.9251 + 0.7771PCULME\$ + 0.0339PCULME\$(-1)
                    (1.90)   (16.92)   (0.77)

$\bar{R}^2$ = 0.971
D.W. = 2.86
SEE = 2.5389
Period of fit = 1960–1976

Production of Copper, Gran Mineria

CUPRODGM = 200.2746 + 0.6394PCUGM/COST(-1)
                      (1.55)   (1.30)

                              + 0.3763CUPRODGM(-1) + 10.2545TREND
                                 (1.52)   (2.79)

$\bar{R}^2$ = 0.717
D.W. = 1.64
SEE = .40.989
Period of fit = 1960–1975

Production of Copper, Medium- and Small-Scale Mining

LCUPRODPMM = 0.5380 + 0.1404LPPMM/PGDP + 0.7867LCUPRODPMM(-1)
                      (1.52)   (1.43)   (12.10)

$\bar{R}^2$ = 0.927
D.W. = 2.24
SEE = .83869E-01
Period of fit = 1960–1976

Total Production of Copper, 000 M.T.

CUPRODTOT = CUPRODGM + CUPRODPMM

Real Total Intermediate Inputs, Gran Mineria

INTGMTOTR = INCOEFTOT · CUPRODGM

Real Domestic Intermediate Inputs, Gran Mineria

INTGMDOMR = INCOEFDOM · CUPRODGM

Current Value Added, Gran Mineria

X2GMN = CUPRODGMN − INTGMTOTN

Real Value Added, Gran Mineria

X2GMR = X2GMN/[(PECUGM$/33.784) × (REXCB/.003128)]

Nominal Wage, Gran Mineria

$$LWGMN = \underset{(-2.81)}{-7.4223} + \underset{(-28.09)}{-0.7742 LPGDP} + \underset{(17.44)}{0.5133 DUM71/72}$$

$$+ \underset{(3.55)}{0.1667\ LOFPGMN(-1)}$$

$\bar{R}^2$ = 0.996
D.W. = 1.42
SEE = .17014
Period of fit = 1961–1976

Number of Persons Employed, Gran Mineria

$$NPEGM = \underset{(0.79)}{1080.7979} + \underset{(1.48)}{0.2400 NPEGM^*} + \underset{(3.99)}{0.7379 NPEGM(-1)}$$

$\bar{R}^2$ = 0.952
D.W. = 0.87
SEE = 1213.5
Period of fit = 1961–1976

Demand for Labor, Gran Mineria

NPEGM* = CUPRODGM/(Q/LGMEST)

Output Per Worker, Gran Mineria (Q/LGMEST)

$$LQ/LGM = \underset{(-.28)}{-0.2745} + \underset{(-0.28)}{0.3397LK/LGM} + \underset{(1.40)}{0.0791LW/PGM} - \underset{(-3.61)}{0.1871DUM71/73}$$

$\bar{R}^2 = 0.721$
$D.W. = 2.00$
$SEE = .64805E\text{-}01$
Period of fit = 1961-1975

Real Gross Investment, Gran Mineria

$$LIBGMR = \underset{(-3.64)}{-22.5746} + \underset{(4.24)}{3.7097LPCUGM/PGDP(-1)}$$

$$+ \underset{(0.51)}{0.6269LPCUGM/PGDP(-3)} - \underset{(-2.32)}{1.6909DUM71/73}$$

$\bar{R}^2 = 0.694$
$D.W. = 2.88$
$SEE = .78241$
Period of fit = 1963-1976

Current Total Production Costs, Gran Mineria in Dollars

$$PRODCOSTGMN\$ = (WBILLGMN\$ + TBMINTGMN\$$$

$$+ INTGMDOMN\$)/CUPRODGM$$

Foreign Trade Sector

Real Imports of Consumption Goods (Except Food), f.o.b.

$$TBMCR = \underset{(1.84)}{0.1332} + \underset{(0.88)}{0.0038CER} + \underset{(2.48)}{0.0581BOPRSVR(-1)} + \underset{(4.40)}{0.1610DUM72/73}$$

$$+ \underset{(5.74)}{0.2487DUM74}$$

$\bar{R}^2 = 0.866$
$D.W. = 1.08$
$SEE = .32157E\text{-}01$
Period of fit = 1961-1976

Real Imports of Food, f.o.b.

TBMFOODR = -0.1310 - 0.6301DEVX1AGR + 0.0373CER
         (-0.58)   (-2.02)   (2.82)

                + 0.4369TBMFOODR(-1) - 0.0017PMFOOD/PX1AG
                  (1.84)   (-1.42)

$\bar{R}^2 = 0.795$
D.W. = 1.74
SEE = .10718
Period of fit = 1961–1976

Real Imports of Capital Goods, Gran Mineria, f.o.b.

LTBMKGMR = -0.3903 + 1.0553LIBGMR - 0.0594TREND
         (-0.98)   (10.60)   (-2.60)

                + 1.0200DUM72/73
                  (2.61)

$\bar{R}^2 = 0.948$
D.W. = 2.01
SEE = .25991
Period of fit = 1960–1976

Real Imports of Capital Goods, Non-Gran Mineria, f.o.b.

TBMKNGMR = MKCOEF × IBNGMR

Real Imports of Intermediate Goods, Gran Mineria, f.o.b.

TBMINTGMR = 0.85INTGMFGNR

Real Imports of Fuel, f.o.b.

TBMFUELR = -0.0669 + 0.0202X2R - 0.0001PMFUEL/PGDP
         (-1.52)   (3.99)   (-1.41)

                + 0.4107TBMFUELR(-1)
                  (2.83)

$\bar{R}^2 = 0.637$
D.W. = 2.46
SEE = .23866E-01
Period of fit = 1961–1976

Real Imports of Intermediate Goods, Non-Gran Mineria, f.o.b.

TBMINOTR = –0.4676 + 0.0645GDPNGMR
            (-2.82)   (7.65)

$\bar{R}^2$ = 0.793
D.W. = 1.05
SEE = .10670
Period of fit = 1960-1975

Implicit Price Deflator for Imports of Goods, in Dollars

PBMG$ = (TBMGN$/TBMGR$) × 100.

Implicit Price Deflator for Imports of Goods, in Pesos

PBMG = (PBMG$ × REXCB)/.003128

Real Exports of Agricultural Goods, f.o.b.

TBEAGR = 0.0274 + 0.0005PEAG/PX1AG + 0.5364TBEAGR(-1)
          (1.87)   (1.83)   (3.01)

                – 0.0230DUM71/73
                  (-3.77)

$\bar{R}^2$ = 0.640
D.W. = 2.20
SEE = .93533E-01
Period of fit = 1961-1976

Real Exports of Other Goods (Excluding Copper and Agriculture), f.o.b.

TBEOTR = 0.2527 + 0.0014PRODEV + 0.4023TBEOTR(-1)
          (2.02)   (1.18)   (2.04)

                – 0.1627DUM72/73
                  (-2.47)

$\bar{R}^2$ = 0.424
D.W. = 1.96
SEE = .77500E-01
Period of fit = 1961-1976

Real Imports of Goods, f.o.b.

$$TBMGR = TBMCR + TBMFOODR + TBMKGMR + TBMKNGMR$$
$$+ TBMINTGMR + TBMFUELR + TBMINOTR$$

Current Imports of Goods, in Dollars, f.o.b.

$$TBMGN\$ = TBMCN\$ + TBMFOODN\$ + TBMKGMN\$ + TBMKNGMN\$$$
$$+ TBMINTGMN\$ + TBMFUELN\$ + TBMINOTN\$$$

Current Imports of Goods, f.o.b.

$$TBMGN = TBMGN\$ \times REXCB$$

Current Exports of Goods, in Dollars, f.o.b.

$$TBEGN\$ = TBEGN\$ + TBECUGMN\$ + TBECUPMMN\$ + TBEOTN\$$$

Current Exports of Goods, f.o.b.

$$TBEGN = TBEGN\$ \times REXCB$$

Implicit Price Deflator for Exports of Goods, in Pesos

$$PEG = TBEGSN/TBEGSR$$

Current Net Nonfactor Services, in Dollars

$$TBNFSN\$ = TBENFSN\$ - TBMNFSN\$$$

Current Balance of Trade, in Dollars

$$TBBGN\$ = TBEGN\$ + TBEGOLDN\$ - TBMGN\$$$

Current Interest Payments on Medium- and Long-Term Debt, in Dollars

$$\text{BOPINTDBN\$} = \underset{(-1.04)}{-29.2818} + \underset{(5.86)}{0.0667}\text{BOPEXTDBMLN\$}(-1)$$

$$\underset{(-2.99)}{-80.1324}\text{DUM71/73}$$

$\bar{R}^2 = 0.818$
D.W. = 1.46
SEE = .34474
RHO(1): 0.355
Period of fit = 1962–1976

Current Net Factor Payments, in Dollars

TBFN$ = TBFDIN$ – BOPINTDBN$

Current-Account Balance, in Dollars

TBCABALN$ = TBBGN$ + TBNFSN$ + TBFN$ + TBURTRN$ + TBSDRN$

Long-Term Capital, Net in Dollars

TBCAPAUTON$ = TBDIN$ + TBMLTLOANN$

Medium- and Long-Term Loans, Net in Dollars

TBMLTLOANN$ = TBMLTLOANDN$ + TBMLTLAMOR$

Balance of Payments Surplus (+) or Deficit (–), in Dollars

BOPBN$ = TBCABALN$ + TBDIN$ + TBMLTLOANN$ + TBERRN$

Net International Reserves, in Dollars

BOPRSVN$ = BOPRSVN$(-1) + BOPBN$

Net Outstanding External Debt, in Dollars

BOPEXTDBMLN\$ = BOPEXTDBMLN\$(-1) + TBMLTLOANN\$

$$- \text{BOPINTDBN\$} + \text{SDEXTDB}$$

Price Sector

Price Deflator, Gross Domestic Product

$$\ln \text{PGDP} = \underset{(1.97)}{0.1635} \ln \text{WULC123} + \underset{(15.91)}{0.6666} \ln \text{PBMG} + \underset{(3.11)}{0.1494} \ln \text{FM1/Y}$$

$$+ \underset{(9.74)}{0.4454} \ln \text{GDPNGMCAPUT}$$

$\bar{R}^2 = 1.000$
$D.W. = 2.35$
$SEE = .42569\text{E-}01$
Period of fit = 1960–1976

Price Deflator, Agriculture

$$\text{LPX1AG} = \underset{(-0.62)}{-0.0466} + \underset{(9.29)}{0.8087\text{LPGDP}} - \underset{(-2.23)}{1.2629\text{LDEVX1AGR}}$$

$$+ \underset{(1.83)}{0.1886\text{LPGDP}(-1)}$$

$\bar{R}^2 = 0.999$
$D.W. = 1.72$
$SEE = .99859\text{E-}01$
Period of fit = 1961–1976

Price Deflator, Manufacturing

$$\ln \text{PX2MFG} = \underset{(0.07)}{0.0022} + \underset{(30.39)}{0.9770} \ln \text{PGDP} + \underset{(0.78)}{0.0298} \ln \text{PGDP}(-1)$$

$\bar{R}^2 = 1.000$
$D.W. = 1.59$
$SEE = .50470\text{E-}01$
Period of fit = 1960–1976

Price Deflator, Construction

$\ln$ PX2CONT = 0.0038 + 0.8661 $\ln$ PGDP + 0.1245 $\ln$ PGDP(-1)
                    (0.04)   (11.80)   (1.48)

$\bar{R}^2 = 0.999$
$D.W. = 1.59$
$SEE = .98448E-01$
RHO(1):0.500
Period of fit = 1960-1976

Price Deflator, Public Utilities

$\ln$ PX2UT = -0.1959 + 0.2750 $\ln$ PGDP + 0.7903 $\ln$ PGDP(-1)
                  (-1.45)   (2.12)   (5.12)

$\bar{R}^2 = 0.995$
$D.W. = 1.41$
$SEE = .20401$
Period of fit = 1960-1976

Price Deflator, Service Sector

$\ln$ PX3 = -0.0292 + 1.0116 $\ln$ PGDP
                (-1.44)   (349.44)

$\bar{R}^2 = 1.000$
$D.W. = 1.25$
$SEE = .35398E-01$
Period of fit = 1960-1976

Price Deflator, Government-Consumption Expenditures

$\ln$ PCG = 0.0894 + 0.9878 $\ln$ PGDP
                (0.89)   (75.94)

$\bar{R}^2 = 0.999$
$D.W. = 1.79$
$SEE = .85245E-01$
Period of fit = 1960-1976

Price Deflator, Private-Consumption Expenditures

LPCP = 0.0328 + 1.0065LPGDP − 0.0754DUM71/73
         (−1.38) (295.87)   (−2.85)

$\bar{R}^2$ = 1.000
D.W. = 1.05
SEE = .41207E–01
Period of fit = 1960–1976

Price Deflator, Gross Investment

ln PIB = 0.0760 + 1.0860 ln PGDP − 0.1116 ln PGDP(−1)
          (3.00)   (44.54)   (−3.85)

$\bar{R}^2$ = 1.000
D.W. = 2.12
SEE = .38273E–01
Period of fit = 1960–1976

Price Deflator, Inventory Investment

ln PICH = −0.2080 + 1.0365 ln PGDP
           (−2.13)   (74.45)

$\bar{R}^2$ = 0.997
D.W. = 1.57
SEE = .17024
Period of fit = 1960–1976

Price Deflator, Capital-Consumption Allowance

ln PCCA = 0.0211 + 0.9931 ln PGDP
           (0.60)   (198.90)

$\bar{R}^2$ = 1.000
D.W. = 1.72
SEE = .61054E–01
Period of fit = 1960–1976

Price Deflator, Indirect Taxes

ln PTXIND = 0.0243 + 0.9985 ln PGDP
　　　　　　(1.66)　(549.88)

$\bar{R}^2$ = 1.000
D.W. = 1.71
SEE = .10158E–01
RHO(1):0.725
Period of fit = 1960–1976

Price Deflator, Transfers and Subsidies

ln PTRSUB = 0.0239 + 0.9985 ln PGDP
　　　　　　(1.69)　(566.10)

$\bar{R}^2$ = 1.000
D.W. = 1.73
SEE = .10081E–01
RHO: 0.711
Period of fit = 1960–1976

Government Sector

Current Indirect Taxes

TXINDN = TXINDRATE × X123NTN

Current Direct Corporate Taxes

LTXDCORPN = –2.3493 + 0.8098LOFPN + 0.1876LOFPN(–1)
　　　　　　　(–35.38)　(8.45)　　　　(1.64)

　　　　　　　　　 – 0.6167DUM71/73
　　　　　　　　　　(–5.67)

$\bar{R}^2$ = 0.970
D.W. = 1.17
SEE = .34945E–01
Period of fit = 1960–1976

Current Direct Income Tax

LTXINCOMN = -4.1468 + 0.9455LYPNTRNAV + 0.1123TREND
                (-33.16)   (20.92)   (3.90)

$\bar{R}^2 = 0.997$
D.W. = 1.57
SEE = .19666
Period of fit = 1961–1976

Current Employers' Social Security Contributions

ln ESCOTN = -1.6997 + 1.0585 ln WGSTOTNAV
               (-40.36) (116.55)

$\bar{R}^2 = 0.999$
D.W. = 1.90
SEE = .10445
Period of fit = 1961–1976

Current Employees' Social Security Contributions

LWESCN = -2.5986 + 0.9804LWGSTOTNAV + 0.0321TREND
             (-67.23)   (76.30)   (4.00)

$\bar{R}^2 = 1.000$
D.W. = 1.44
SEE = .52242E-01
Period of fit = 1961–1976

Current Government Revenues

GVRVN = GVRVPFPN - GVINTERN + TXINDN + TXDCORPN + TXINCOMN
                       + WESCN + ESCTOTN + TXTRINDVN + TXTRFGNN

Current Government Expenditures

GVEXPN = CEGN + TRSUBGN + TRINDVN + TRFGNN

Current Government Savings

GVSVN = GVRVN - GVEXPN

Real Government Savings

GVSVR = GVSVN/(PGDP/100)

Monetary Sector

Legal Reserve Requirement

RRCB = 8.8382 + 5.3639LPGDP + 44.0307DUM71/73
          (1.23)   (5.19)   (5.48)

$\bar{R}^2$ = 0.800
D.W. = 1.73
SEE = 12.522
Period of fit = 1960–1976

Monetary Multiplier

FMPLY = 1.97508 - 0.0109RRCB + 0.5535DUM73
          (28.36)   (-7.78)   (3.41)

$\bar{R}^2$ = 0.813
D.W. = 1.72
SEE = .10704
Period of fit = 1960–1976

Real Flow of Credit from Central Bank to Government

DFCRCBGR = 3.3450 - 1.7822GVSVNINTR + 7.1187DUM73
              (4.14)   (-2.11)   (2.47)

$\bar{R}^2$ (CORR) = 0.541
D.W. = 0.72
SEE = 2.4312
Period of fit = 1960–1976

*Nominal Credit Outstanding from Central Bank to
the Government*

FCRCBGN = FCRCBGN(-1) + DFCRCBGR · (PGDP/100)

Monetary Base

FMBASE = BOPRSVN + FCRCBGN + FMBASEOTHER

Factors of Production

Real Wage, Non-Gran Mineria

$$\text{W123NGMR} = -1.0801 + 0.5545\text{Q/LNGM} - 0.0013\text{ PGDP*} + 0.5786\text{DUM71/73}$$
$$\quad\quad\quad\quad (-1.54)\quad (5.67)\quad\quad\quad\quad (-4.68)\quad\quad\quad\quad (4.40)$$

$\bar{R}^2 = 0.845$
D.W. = 2.66
SEE = .18714
Period of fit = 1961–1976

Current Wage Bill, Non-Gran Mineria

WBILLNGMN = (W123NGMN × NPENGM)/1000

Current Wage, Non-Gran Mineria

W123NGMN = W123NGMR × (PGDP/100)

*Real Output per Worker, Non-Gran Mineria Estimate
(GDPNGM/NPEEST)*

$$\ln \text{GDPNGM/NPE} = 0.7333 + 0.1741 \ln \text{W/PGDPNGM}$$
$$\quad\quad\quad\quad\quad (5.32)\quad (3.19)$$

$$+ 0.5178 \ln \text{GDPNGM/NPE}(-1)$$
$$(5.40)$$

$\bar{R}^2 = 0.941$
D.W. = 2.53

SEE = .15048E–01
Period of fit = 1961–1971

Employment, Non-Gran Mineria

NPENGM = 22.6667 + 0.5498NPENGM* + 0.4510NPENGM(–1)
(0.15)   (2.84)   (2.59)

$\bar{R}^2$ = 0.961
$D.W.$ = 1.75
SEE = 47.746
Period of fit = 1961–1976

Current Wage Bill, Total

W123N = WBILLNGMN + WBILLGMN

Current Wages and Salaries, Total

WGSTOTN = W123N – ESCTOTN

Unit Labor Cost

WULC123 = W123N/X123R

Current Other Factor Payments, Total

OFPN = YNN + TFBFN – W123N

*Current Other Factor Payments, Total Net of Factor Payments
Abroad and Net of Taxes*

OFPNNTN = OFPNN – (TXRATECORP × OFPNN)

Current Wages and Salaries Net of Taxes

WGSTOTNTN = WGSTOTN – TXINCOMN

Current National Income

YNN = GDPN – TFBFN – TXINDN + TRSUBGN – CCAN

Current Personal Income

YPN = YNN – SAVCORPN – TXDCORPN – GVRVPFPN + GVINTERN

+ TRINDVN + FGNTRPERN – TXTRINDVN

Current Personal Income, Net of Transfers to Individuals

YPNTRN = YPN – TRINDVN

List of Variables

BOPBN:	Balance of payments balance, millions of pesos
BOPBN$:	Balance of payments, surplus or deficit, millions of U.S. dollars
BOPBR:	Balance of payments, surplus or deficit, in real pesos, millions of 1965 pesos
BOPEXTDBMLN$:	Net external debt, medium- and long-term debt, millions of U.S. dollars
BOPINTDBN$:	Interest on medium- and long-term debt, millions of U.S. dollars
BOPRSVN$:	Balance-of-payments level of international reserves, millions of pesos
C:	User cost of capital, percent
CCAGMN:	Depreciation: Gran Mineria, millions of pesos
CCAN:	Current depreciation allowances, millions of pesos
CCAR:	Real depreciation allowances, millions of 1965 pesos
CEGN:	Government consumption, millions of pesos
CEGR:	Real government consumption, millions of 1965 pesos
CEN:	Current consumption expenditures, millions of pesos
CEPN:	Private consumption, millions of pesos

CEPR:	Real private consumption, millions of 1965 pesos
CER:	Real total consumption expenditures, millions of 1965 pesos
CORPYN:	Corporate gross income, millions of pesos
CORPYR:	Real corporate gross income, millions of 1965 pesos
CUEXGM:	Exports of copper (volume): Gran Mineria, 000 M.T.
CUEXPMM:	Exports of copper (volume): pequena y mediana minerias, 000 M.T.
CUINV:	Copper inventories, 000 M.T.
CUPRODGM:	Production of copper: Gran Mineria, 000 M.T.
CUPRODPMM:	Production of copper: pequena y mediana minerias, 000 M.T.
CUPRODTOT:	Production of copper: Chile, 000 M.T. refined
CUSADOM:	Domestic sales of copper, 000 M.T.
DEPRC:	Real rate of depreciation, percent
DFCRCBGN:	FCRCBGN - FCRCBGN(-1), millions of pesos
EGSR:	Real exports of goods and services, millions of 1965 pesos
ESCTOTN:	Total employers' Social Security contributions, millions of pesos
FCRCBGN:	Domestic credit Central Bank to public sector, millions of pesos
FGNTRPERN:	Foreign transfers to persons, millions of pesos
FMBASE:	Monetary base, millions of pesos
FMBASEOTHER:	Other net assets: Central Bank, millions of pesos
FMPLY:	Money multiplier, unit
FM1:	Total money in circulation: M1, millions of pesos
FM1/Y*:	Percentage change FM1/Y, percent
GDPN:	Current gross domestic product, millions of pesos
GDPNGM/NPE:	Output per worker, noncopper, millions of 1965 pesos
GDPNGMCAP:	Capacity output: non-Gran Mineria, millions of 1965 pesos
GDPNGMCAPUT:	Rate of capacity utilization: non-Gran Mineria GDP, percent
GDPNGMN:	Current gross domestic product excluding Gran Mineria, millions of pesos

GDPNGMR:	Real gross domestic product excluding Gran Mineria, millions of 1965 pesos
GDPR:	Real gross domestic product, millions of 1965 pesos
GNPR:	Real gross national product, millions of 1965 pesos
GVEXPN:	Current government expenditures, millions of pesos
GVINTERN:	General government: interest payments, millions of pesos
GVRVN:	General government: current revenues, millions of pesos
GVRVPFPN:	General government: income from public firms and properties, millions of pesos
GVRVR:	Real government revenues, millions of 1965 pesos
GVSVN:	General government: current savings, millions of pesos
GVSVNINTN:	Public sector: current savings net of interest payments, millions of pesos
GVSVNINTR:	Public sector: real current savings net of interest payments, millions 1965 pesos
GVSVR:	Real government savings (deflated by PIB), millions of 1965 pesos
IBGMR:	Real gross investment: Gran Mineria, millions of 1965 pesos
IBN:	Current gross fixed investment, millions of pesos
IBNGMN:	Current gross investment excluding Gran Mineria, millions of pesos
IBNGMR:	Real gross investment excluding Gran Mineria, millions of 1965 pesos
IBR:	Real gross fixed investment, millions of 1965 pesos
ICHN:	Current change in inventories, millions of pesos
ICHR:	Real change in inventories, millions of 1965 pesos
INCOEFGMDOMM:	Gran Mineria: coefficient of domestic intermediate inputs, fraction
INCOEFGMTOT:	Gran Mineria: coefficient of intermediate inputs, fraction
INTERPERN:	Interest paid by persons, millions of pesos
INTGMDOMN:	Domestic intermediate inputs to Gran Mineria, millions of pesos
INTGMDOMN$:	Current domestic inputs to Gran Mineria, millions of U.S. dollars

INTGMDOMR: Real domestic intermediate inputs to Gran Mineria, millions of 1965 pesos

INTGMFGNN: Imported intermediate inputs to Gran Mineria, millions of pesos

INTGMFGNR: Real FGN intermediate inputs to Gran Mineria, millions of 1965 pesos

INTGMTOTN: Intermediate inputs to Gran Mineria, millions of pesos

INTGMTOTR: Total real intermediate inputs to Gran Mineria, millions of 1965 pesos

INV: All stock of inventories, millions of 1965 pesos

KGMR: Real stock of capital: Gran Mineria, millions of 1965 pesos

KNGMR: Real stock of capital: non-Gran Mineria, millions of 1965 pesos

KR: Real stock of capital, millions of 1965 pesos

KUTNGM: Utilized capital stock: non-Gran Mineria, millions of 1965 pesos

LCUPRODGM: logCUPRODGM, tran

LDEVX1AGR: Difference of logs of X1AGR and X1AGR, long-run trend, 1965 pesos

LW/PGM: log(W/PGM), tran

MCOEFK: Import coefficient W.R.T. investment: non-Gran Mineria, fraction

MGSR: Real imports of goods and services, millions of 1965 pesos

MKC/GDP: Imports of K and C goods as percent of GDPR, percent

NP: Total population, Chile, thousands of persons

NPACT: Economically active population, Chile, thousands of persons

NPEGM: Number of persons employed: Gran Mineria, thousands of persons

NPEGMMAX: Trend through peaks of full employment: Gran Mineria, thousands of persons

NPENGM: Number of persons employed excluding Gran Mineria, thousands of persons

NPENGMMAX: Noncopper maximum employment, thousands of workers

NRU: Rate of unemployment, Chile, percent

NRUGM: Rate of unemployment, Gran Mineria, percent

OFPGMN: Other factor payments: Gran Mineria, millions of pesos

OFPN: Other factor payments in current pesos, millions of pesos

OFPNGMN: Other factor payments excluding Gran Mineria, millions of pesos

OFPNN: Other factor payments excluding FGN factor payments, millions of pesos

OFPNNTR/N: Real per capita OFPNNTN, millions of 1965 pesos

PBMG: Implicit deflator, imports of goods, balance-of-payments account, 1965 = 100

PBMG$:   Price index, imports of goods, U.S. dollars, 1965 = 100

PCCA: Implicit deflator, capital-consumption allowance, 1965 = 100

PCG: Implicit deflator, government-consumption, 1965 = 100

PCP: Implicit deflator, private-consumption expenditures, 1965 = 100

PCUGM/COST: Price of exports divided by cost of production: Gran Mineria, 1965 = 100

PCUGM/PGDP: Price of copper exports divided by PGDP, pesos

PCULME$:   Copper price: London Metal Exchange, US¢/lb

PEAG$:   Price index of agricultural exports, 1965 = 100

PECUGM$:   Average price of copper exports: Gran Mineria, US¢/lb

PECUPMM$:   Average price of copper exports: pequena y mediana mineria, US¢/lb

PEG: Price deflator, exports of goods, 1965 = 100

PEOT: Export price index excluding copper and agriculture, 1965 = 100

PGDP: Implicit deflator, gross domestic product, 1965 = 100

PGDP*:   Rate of change of PGDP, percent

PIB: Implicit deflator, gross fixed investment, 1965 = 100

PICH: Implicit deflator, change in inventories, 1965 = 100

PMCAPITAL: Import price index of capital goods, 1965 = 100

PMCONSUMER: Import price index of consumer goods, 1965 = 100

PMFOOD: Import price index of food, 1965 = 100

PMFUELS: Import price index of fuels, 1965 = 100

PMICOPPER: Import price index intermediates, copper and nonfuel, 1965 = 100

PMINOT: Import price index, intermediate other, 1965 = 100

PMKNGM: Import price index, capital goods, non-Gran Mineria, 1965 = 100

PRODCOSTGMN$:   Cost of production, Gran Mineria, millions of U.S. dollars

PRODEV: Index of industrial production, OECD countries, 1970 = 100

PRODFUELR: Real value of production of crude petroleum and coal, millions of 1965 pesos

PTRSUB: Implicit deflator, government subsidies, 1965 = 100

PTXIND: Implicit deflator, indirect taxes, 1965 = 100

PX1AG: Implicit deflator, agricultural sector, 1965 = 100

PX2CONT: Implicit deflator, construction, 1965 = 100

PX2MFG: Implicit deflator, manufacturing, 1965 = 100

PX2MINNGM: Implicit deflator, non-Gran Mineria mining, 1965 = 100

PX2UT: Implicit deflator, utilities, 1965 = 100

PX3:   Implicit deflator, tertiary sector, 1965 = 100

Q/LGM: Output per worker, Gran Mineria, 000 M.T./person

REXCB: Exchange rate, commercial banks, pesos/U.S. dollar

REXGM: Exchange rate, Gran Mineria, pesos/U.S. dollar

RRCB: Reserve-requirement ratio, percent

RUSBLTN$:   U.S. government bond yield, long-term, percent

S: Gross domestic sales, millions of 1965 pesos

SAVCORPN: Savings by corporations, millions of pesos

TBBGN$:   Balance-of-payments account, current balance of goods, millions of U.S. dollars

TBCABALN$:   Balance of payments, current account balance, millions of U.S. dollars

TBDIN$:   Balance of payments, foreign direct investment, millions of U.S. dollars

TBEAGN$:   Exports of agricultural products (f.o.b.), millions of U.S. dollars

TBEAGR: Real exports of agricultural goods (Balance-of-payments account), millions of 1965 pesos

TBECUGMN$:   Exports of copper, Gran Mineria (f.o.b.), millions of U.S. dollars

TBECUGMR: Real exports of copper, Gran Mineria, millions of 1965 pesos

TBECUPMMN$:   Exports of copper: pequena y mediana mineria, (f.o.b.), millions of U.S. dollars

TBECUPMMR: Real exports of copper: pequena y mediana mineria, millions of 1965 pesos

TBEGN$:   Balance-of-payments account, current exports of goods, millions of U.S. dollars

TBEGOLDN$:   Exports of nonmonetary gold, millions of U.S. dollars

TBEGR: Balance-of-payments account, real exports of goods, millions of 1965 pesos

TBENFSN$:   Balance-of-payments account, current exports of nonfactor services, millions of U.S. dollars

TBEOTN$:   Current exports excluding copper and agriculture, millions of U.S. dollars

TBERRN$:   Balance-of-payments errors and omissions, millions of U.S. dollars

TBFDIN$:   Balance-of-payments direct investment income, millions of U.S. dollars

TBFN$:   Balance of payments, net factor payments abroad, millions of U.S. dollars

TBMCN$:   Imports of consumption goods (excluding food, f.o.b.), millions of U.S. dollars

TBMCR: Real imports of consumption goods (excluding food, f.o.b.), millions of 1965 pesos

TBMFOODN$:   Imports of food (f.o.b.), millions of U.S. dollars

TBMFOODR: Real imports of food (f.o.b.), millions of 1965 pesos

TBMFUELN$:   Imports of fuels, millions of U.S. dollars

TBMFUELR: Real imports of fuel, millions of 1965 pesos

TBMGN$:   Balance-of-payments account, current imports of goods, millions of U.S. dollars

TBMGR: Real imports of goods (f.o.b., Balance-of-payments account), millions of 1965 pesos

TBMGSR: Real imports of goods and services (Balance-of-payments account), millions of 1965 pesos

TBMINOTN$:   Current imports of other intermediates (f.o.b.), millions of U.S. dollars

TBMINOTR: Real imports of other intermediates, millions of 1965 pesos

TBMINTGMN$:   Imports of intermediates, Gran Mineria (f.o.b.), millions of U.S. dollars

TBMINTGMR: Real imports of intermediates, Gran Mineria, millions of 1965 pesos

TBMINTN$:   Imports of intermediate goods (excluding food, f.o.b.), millions of U.S. dollars

TBMINTR: Real imports of intermediate goods, millions of 1965 pesos

TBMKGMN$:   Current imports of capital goods, Gran Mineria, millions of U.S. dollars

TBMKGMR: Real imports of capital goods, Gran Mineria, millions of 1965 pesos

TBMKNGMN$:   Current imports of capital excluding Gran Mineria, millions of U.S. dollars

TBMKNGMR: Real imports of capital excluding Gran Mineria, millions of 1965 pesos

TBMKR: Real imports of capital goods (f.o.b.), millions of 1965 pesos

TBMLTLAMOR$:   Balance-of-payments medium- and long-term loans, amortization, millions of U.S. dollars

TBMLTLOANDN$:   Balance-of-payments medium- and long-term loans, disbursements, millions of U.S. dollars

TBMLTLOANN$:   Balance-of-payments medium- and long-term loans (net), millions of U.S. dollars

TBMNFSN$:   Balance-of-payments account, current imports of nonfactor services, millions of U.S. dollars

TBMNFSR: Real imports of services, millions of 1965 pesos

TBNFSN$:   Balance-of-payments, current net nónfactor payments, millions of U.S. dollars

TBSDRN$:   Allocation of SDR's, millions of U.S. dollars

TBURTRN$:   Unrequited transfers, millions of U.S. dollars

TFBFN: Current net payments to foreign factors in NIA, millions of pesos

TFBFR: Real net payments to foreign factors in NIA, millions of 1965 pesos

TREND: Time trend, 1960 = 1, tran

TRFGNN: General government, current transfers abroad, millions of pesos

TRINDVN: General government, current transfers to individuals, millions of pesos

TRSUBGN: Government subsidies, millions of pesos

TXDCORPN: Direct taxes on corporations, millions of pesos

TXINCOMN: General government, personal income tax, millions of pesos

TXINDN: Indirect taxes, millions of pesos

TXINDR: Real indirect taxes, millions of 1965 pesos

TXINDRATE: Tax rate: indirect taxes, percent

TXRATECORP: Average tax rate on corporations, percent

TXTRFGNN: General government, current transfers received from abroad, millions of pesos

TXTRINDVN: General government, other current transfers from individuals, millions of pesos

W/NPENGM: Wage bill per worker, noncopper, pesos/worker

W/PGDPNGM: Real wage rate, 1965 pesos

W/PGM: Real wage, Gran Mineria, thousands of 1965 pesos

WBILLGMN: Wage bill, Gran Mineria, millions of pesos

WESCN: General government, workers' and employees' Social Security contributions, millions of pesos

WGMN: Wage per worker, Gran Mineria, thousands of pesos

WGSTOTN: Total wages and salaries, millions of pesos

WGSTOTNTN:	Current wages and salaries net of income tax, millions of pesos
WGSTOTNTR/N:	Real per capita WGSTOTNTN, millions of 1965 pesos
WULC123:	Unit labor cost, overall economy, pesos per output
WULC123*:	Percentage change WULC123, percent
W123N:	Total wage bill, millions of pesos
W123NGMN:	Wage per worker, non-Gran Mineria, thousands of pesos/person
X1AGN:	Current value added, agriculture, fishing, forestry, millions of pesos
X1AGR:	Real value added, agriculture, fishing, forestry, millions of 1965 pesos
X1AGRLT:	Long-run trend, real value added, agriculture, millions of 1965 pesos
X12NGMR:	Value added, primary and secondary sectors excluding Gran Mineria, millions of 1965 pesos
X12R:	Real value added, primary and secondary sectors, millions of 1965 pesos
X123N:	Total current value added, millions of pesos
X123NTM:	GDPN net of indirect taxes, millions of pesos
X123R:	Total real value added, millions of 1965 pesos
X2CONTN:	Current value added, construction, millions of pesos
X2CONTR:	Real value added, construction, millions of 1965 pesos
X2GMN:	Current value added, Gran Mineria, millions of pesos
X2GMR:	Real value added, Gran Mineria, millions of 1965 pesos
X2MFGN:	Current value added, manufacturing, millions of pesos
X2MFGR:	Real value added, manufacturing, millions of 1965 pesos
X2MINNGMN:	Current value added, mining, excluding Gran Mineria, millions of pesos
X2MINNGMR:	Real value added, mining, excluding Gran Mineria, millions of 1965 pesos
X2N:	Current value added, secondary sector, millions of pesos
X2R:	Real value added, secondary sector, millions of 1965 pesos

X2UTN:	Current value added, utilities, millions of pesos
X2UTR:	Real value added, public utilities, millions of 1965 pesos
X3N:	Current value added, tertiary sector, millions of pesos
X3R:	Real value added, tertiary sector, millions of 1965 pesos
YNN:	National income, millions of pesos
YPN:	Personal income, millions of pesos
YPNTRN:	Personal income net of transfers to individuals, millions of pesos

References

Adams, F.G., and J.R. Behrman (Eds.). 1978. *Econometric Modeling of World Commodity Policy.* Lexington, Mass.: Lexington Books, D.C. Heath and Co.
——. 1981. *Commodity Exports and Economic Development: The Commodity Problem, Goal Attainment and Policy in Developing Countries.* Lexington, Lexington Books, D.C. Heath and Co.
Adams, F.G., and R.A. Roldan. 1978. "Econometric Studies of the Impact of Primary Commodity Markets on Economic Development in Latin America." Paper prepared for the NBER conference on commodity markets, models, and policies in Latin America, Lima, Peru.
Ando, A., F. Modigliani, R. Rasche, and S. Turnovsky. 1974. "On the Role of Expectations of Price and Technological Change in an Investment Function." *International Economic Review* 15, June.
Arbildua, B., and R. Luders. 1968. "Una Evaluacion Comparada De Tres Programas Anti-Inflacionarios En Chile: Una Decada De Historia Monetaria: 1956-1966." *Cuadernos de Economia* 14.
Ballesteros, M. 1965. "Desarrollo Agricola Chileno 1910-1955." *Cuadernos de Economia* 5.
Banco Central de Chile. 1965-1975. *Balanza De Pagos de Chile.* Santiago.
Banco Central de Chile. 1960-1978. *Boletin Mensual.* Santiago.
Banco Central de Chile. 1968. *Estudios Monetarios.* Santiago de Chile.
Banco Central de Chile. 1970. *Estudios Monetarios II.* Santiago de Chile.
Banco Central de Chile. 1978. "Programa Economico 1978." *Boletin Mensual,* February.
Banks, F. 1974. *The World Copper Market: An Economic Analysis.* Cambridge, Mass.: Ballinger Publishing Co.
Barria, J. 1974. "Organizacion Y Politicas laborales En La Gran Mineria Del Cobre." in: R.F. Ffrench-Davis and E. Tironi (eds). *El Cobre En el Desarrollo Nacional.* Santiago: Ediciones Nueva Universidad, Universidad Catolica de Chile.
Behrman, J.R. 1972. "Sectoral Investment Determination in Developing Economy." *American Economic Review.* 62.
——. 1973. "Price Determination in an Inflationary Economy, the Dynamics of Chilean Inflation Revisited." in: R. Eckaus and P.N. Rosenstein-Rodan (eds). *Analysis of Development Problems: Studies of the Chilean Economy.* Amsterdam: North Holland.
——. 1977. *Macroeconomic Policy in a Developing Country: The Chilean Experience.* Amsterdam: North Holland.
—— 1968. "International Commodity Market Structures and the Theory Underlying International Commodity Market Models." in F.G. Adams and

J.R. Behrman (eds). *Econometric Modeling of World Commodity Policy.* Lexington, Mass.: Lexington Books, D.C. Heath and Co.

Behrman, J.R. and L.R. Klein. 1970. "Econometric Growth Models For The Developing Economy." in: M.F.G. Scott and J.N. Wolf (eds). *Induction, Growth and Trade.* Oxford: Oxford University Press.

Bitar, S. 1970. "Politicas De Desarrollo Industrial." *Cuadernos de Economia* 22.

———. 1974. "Efectos De Las Areas De Propiedad Social y Mixta En La Industria Chilena." *El Trimestre Economico* 163.

Bitar, S., and C. Bradford. 1967. "Problematica De La Politica En Chile: La Compatibilizacion de las Metas Economicas." *Cuadernos de Economia* 13.

Branson, W.H. 1972. *Macroeconomic Theory and Policy.* New York: Harper and Row.

Business International Corporation. 1975. *Chile after Allende.* New York: Business International Corp.

Cabezon, Pedro. 1971. "Antecedentes Historicos De Las Importaciones y De La Politica Comercial En Chile." *Cuadernos de Economia* 25.

Castro, S. 1972. "Analisis De la Politica Salarial En El Periodo 1960-1961." *Cuadernos de Economia* 28.

Cauas, J. 1970. "Stabilization Policy—The Chilean Case." *Journal of Political Economy* 78, July-August.

Contador, C.R. 1974. "Desarrollo De Las Instituciones Financieras No Bancarias: Ayuda o Impasse Para La Politica Monetaria?" *Cuadernos de Economia* 33.

Copper Studies, Inc. 1978. *Copper Studies.* New York.

Corbo Lioi, V. 1974. *Inflation in Developing Countries: An Econometric Study of Chilean Inflation.* Amsterdam: North-Holland.

CORFO. 1976-1979. *Chile Economic News.* New York: CORFO

Corporacion del Cobre. 1974. "Tendencias De Los Tipos De Cambio Periodo Enero 1970-Abril 1974." Santiago.

———. 1975. *CODELCO: Anuario.* Santiago.

———. 1977. *1ª Memoria Anual 1976.* Santiago.

———. 1978. *2ª Memoria Anual 1977.* Santiago.

Cramer, J.S. 1971. *Empirical Econometrics.* Amsterdam: North-Holland.

Cruzat, M. 1969. "Algunas Ideas Con Respecto Al Mercado De Capitales En Chile." *Cuadernos de Economia* 17.

Cuadra, S., E.R. Fontaine, and R. Luders. 1972. "Algunas Consideraciones En Torno Al Problema De La Distribucion Del Ingreso y El Empleo." *Cuadernos de Economia* 28.

Davis. T. 1966. "Capital y Salarios Reales En La Economia Chilena." *Cuadernos de Economia* 8.

———. 1967. "Ocho Decadas De Inflacion En Chile 1879-1959: Una Interpretacion Politica." *Cuadernos de Economia* 11.

DeCastro, S. 1978. "Exposicion Sobre El Estado De La Hacienda Publica: Presentada for El Ministro De Hacienda Sr. Sergio De Castro Spikula." In Banco Central de Chile, *Boletin Mensual.* Santiago, January.

Eckaus, R., and P. N. Rosenstein-Rodan (Eds.). 1973. *Analysis of Development Problems: Studies of the Chilean Economy.* Amsterdam: North-Holland.

Eckstein, O., and G. Fromm. 1968. "The Price Equation." *American Economic Review* 58, December.

Eisner, R., and R.H. Strotz. 1963. "Determinants of Business Investment." In Commission on Money and Credit, *Impact of Monetary Policy.* Englewood Cliffs, N.J.: Prentice-Hall.

Evans, M.K. 1969. *Macroeconomic Activity.* New York: Harper and Row.

Faundez, J. 1978. "A Decision without a Strategy: Excess Profits in the Nationalization of Copper in Chile." In J. Faundez and S. Picciotto (eds.), *The Nationalisation of Multinationals in Peripheral Economies.* London: Macmillan.

Feldstein, M., and R. Auerbach. 1976. "Inventory Behavior in Durable Goods Manufacturing: The Target Adjustment Model." *Brookings Papers in Economic Activity* 2.

Fisher, F., P. Cootner and M. Baily, 1972. "An Econometric Model of the World Copper Industry." *Bell Journal of Economics and Management Science.* 3.

Fortin, C. 1978. "Law and Economic Coercion as Instruments of International Control: The Nationalisation of Chilean copper." In J. Faundez and S. Picciotto (Eds.), *The Nationalisation of Multinationals in Peripheral Economies.* London: Macmillan.

Ffrench-Davis, R. 1973. *Politicas Economicas En Chile 1952-1970.* Santiago: Ediciones Nueva Universidad, Universidad Catolica de Chile.

———. 1974a. "La Importancia Del Cobre En La Economia Chileana." In R. Ffrench-Davis and E. Tironi (Eds.), *El Cobre En El Desarrollo Nacional.* Santiago: Ediciones Nueva Universidad, Universidad Catolica de Chile.

———. 1974b. "Integracion De La Gran Mineria a la Economia Nacional: El Rol De Las Politicas Economicas." In R. Ffrench-Davis and E. Tironi (Eds.), *El Cobre En El Desarrollo Nacional.* Santiago: Ediciones neuva Universidad, Universidad Catolica de Chile.

———. 1975. "Sustitucion Entre Insumos Nacionales E Importados En la Produccion De Cobre: Una Estimacion Econometrica." Documento No. 46, CIEPLAN, Chile.

Ffrench-Davis, R., and E. Tironi (Eds.). 1974. *El Cobre En El Desarrollo Hacional.* Santiago: Ediciones Nueva Universidad, Universidad Catolica de Chile.

Garces, F. 1968. "Estructura y Cambio De Las Finanzas Publicas De Chile 1950-1967." *Boletin Mensual.* Banco Central de Chile, December.

Garcia, E. 1964. "Inflation in Chile: A Quantitative Analysis." PhD dissertation, MIT, Cambridge, Mass.

Garcia, E., S. Baeza, and S. Perez. 1973. "La Distribucion Del Excedente En Las Empresas Estatales: Una Metodologia General y Su Aplicacion Al Caso De ENDESA." *Cuadernos de Economia* 29.

Hagen, E.E. 1975. *The Economics of Development*. Homewood, Ill.: Irwin.

Harberger, A.C., and M. Selowsky. 1966. "Fuentes Del Crecimiento Chileno." *Cuadernos de Economia* 10.

Johnston, J. 1972. *Econometric Methods*. New York: McGraw-Hill.

Jorgenson, D.W. 1963. "Capital Theory and Investment Behavior." *American Economic Review* 53, May.

Jorgenson, D.W., and C.D. Siebert. 1968. "A Comparison of Alternative Theories of Corporate Investment Behavior." *American Economic Review* 59, September.

Klein, L.R. 1962. *An Introduction to Econometrics*. Englewood Cliffs, N.J.: Prentice-Hall.

Knudsen, O., and A. Parnes. 1975. *Trade Instability and Economic Development*. Lexington, Mass.: Lexington Books, D.C. Heath and Co.

Lessard, D.R. 1977. "Estrategias Financieras Externas Eficientes En Cuanto A Riesgo Para Paises Productores de Bienes Primarios." *Cuadernos de Economia* 42.

Lira, R. 1974. "The Impact of an Export Commodity in a Developing Economy: The Case of the Chilean Copper 1956-1968." Unpublished PhD dissertation, University of Pennsylvania.

Luders R. 1969. "El sistema Tributario Chileno: Algunos Comentarios." *Cuadernos de Economia* 17.

———. 1970. "Una historia Monetaria de Chile: 1925-1958." *Cuadernos de Economia* 20.

MacBean, A. 1966. *Export Instability and Economic Development*. Cambridge, Mass.: Harvard Univ. Press.

McNicol, D. 1975. "The Two Price System in the Copper Industry." *Bell Journal of Economics* 6, Spring.

Maizels, A. 1968. "Review of Export Instability and Economic Development." *American Economic Review* 58, June.

Mamalakis, J. 1976. *The Growth and Structure of the Chilean Economy: From Independence to Allende*. New Haven: Yale Univ. Press.

Manger, J. 1979. "A Review of the Literature Causes, Effects and Other Aspects of Export Instability." Report of Wharton E.F.A., Inc. for AID Project on Primary Commodity Stabilization and Economic Development.

Moran, E. 1974. "Estimacion de la Tasa de Retorno del Capital." *Cuadernos de Economia* 34.

Nisbet, C.T. 1966. "El Mercado de Credito No-Institucional De Chile Rural." *Cuadernos de Economia* 10.

Oficina De Planificacion. 1977. *Cuentas Nacionales de Chile 1960–1975.* Santiago.

Ossa, C. 1964. "La Politica monetaria y La Programacion Del Desarrollo Economico." *Cuadernos de Economia* 3.

Pindyck, R.S. 1978. "Gains to Producers from the Cartelization of Exhaustible Resources." *Review of Economics and Statistics* 60, May.

Pobukadee, J. 1979. "An Econometric Analysis of the World Copper Market." Wharton E.F.A., Inc., Philadelphia.

Priovolos, T. 1981. *Coffee and the Ivory Coast: An Econometric Study.* Lexington, Mass.: Lexington Books, D.C. Heath and Co.

Ramos, J. 1977. "Inflacion Persistente, Inflacion Reprimida E Hiperstanflacion, Lecciones De Inflacion y Estabilizacion En Chile." *Cuadernos de Economia* 43.

Rangarajan, C., and V. Sundararayan. 1976. "Impact of Export Fluctuations on Income: A Cross Country Analysis." *Review of Economics and Statistics* 58, August.

Selowsky, M. 1973. "Cost of Price Stabilization in an Inflationary Economy." *Quarterly Journal of Economics* 87, February.

Tironi, E. 1974. "Planificacion Economica en el Sector Cuprero Nacionalizado." In R.Ffrench-Davis and E. Tironi (Eds.), *El Cobre En El Dessarrollo Nacional.* Ediciones Nueva Universidad, Universidad Catolica de Chile.

———. 1977. "Issues in the Development of Resource-Rich LDCs: Copper in Chile." CIEPLAN, Santiago, November.

———. 1978. "Politicas Nacionales Alternativas De Paises En Desarrollo Respecto Al Comercio De Productos Basicos." Seminario Sobre Areas Prioritarias De Investigacion Economica Internacional y Paises En Desarrollo, Santiago, Chile.

Valdes, A. 1972. "Politica Comercial y Su Efecto Sobre El Comercio Exterior Agricola En Chile: 1945-1965." *Cuadernos de Economia* 28.

———. 1973. "La Transicion Al Socialismo Observaciones Sobre La Agricultura Chilena." *Cuadernos de Economia* 29.

Varas, J.I. 1975. "El Impacto de una Liberalizacion Del Comercio En El Sector Agricola Chileno." *Cuadernos de Economia* 36.

Wachter, S.M. 1976. *Latin American Inflation: The Structuralist-Monetarist Debate.* Lexington, Mass.: Lexington Books, D.C. Heath and Co.

Wallis, K.F. 1973. *Topics in Applied Econometrics.* London: Gray-Mills.

Zorn, S.A. 1978. "Producers Associations and Commodity Markets: The Case of C.I.P.E.C." In F.G. Adams and S.A. Klein (Eds.), *Stabilizing World Commodity Markets.* Lexington: Lexington Books, D.C. Heath and Co.

Index

Index

About the Author

Manuel Lasaga is currently senior economist at Wharton Econometric Forecasting Associates, Inc., where he is director of the Mexican Forecasting Service. He received the B.A. in economics from the University of Maryland, and the M.A. and Ph.D. in economics from the University of Pennsylvania. Dr. Lasaga has worked for the World Bank and was a contributing author for a book on the economy of the Dominican Republic. His recent work has focused on modeling of the financial sector of a developing economy with an emphasis on developing industry models.